I0045044

Anonymous

Standard Atlas of Clinton County, Michigan

Including a Plat Book of the Villages, Cities and Townships of the County

Anonymous

Standard Atlas of Clinton County, Michigan
Including a Plat Book of the Villages, Cities and Townships of the County

ISBN/EAN: 9783337715045

Printed in Europe, USA, Canada, Australia, Japan

Cover: Foto ©berggeist007 / pixelio.de

More available books at **www.hansebooks.com**

STANDARD ATLAS

OF

CLINTON COUNTY

MICHIGAN

INCLUDING

A PLAT BOOK

OF THE

VILLAGES, CITIES AND TOWNSHIPS OF THE COUNTY.

MAP OF THE STATE, UNITED STATES AND WORLD.

Farmers Directory, Reference Business Directory and Departments devoted to General Information.

ANALYSIS OF THE SYSTEM OF U.S. LAND SURVEYS, DIGEST OF THE SYSTEM OF CIVIL GOVERNMENT, ETC. ETC.

Compiled and Published

BY

GEO. A. OGLE & CO.

PUBLISHERS & ENGRAVERS.

134 VAN BUREN ST.

CHICAGO.

1896

TABLE OF CONTENTS.

GENERAL INDEX.

CLINTON COUNTY INDEX.

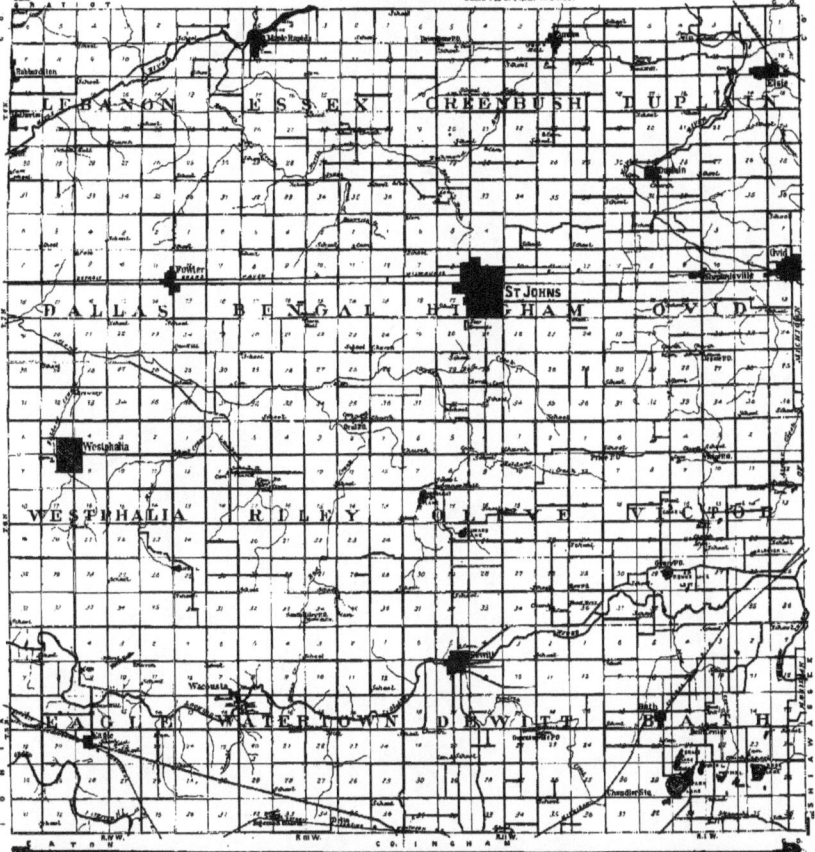

OUTLINE MAP OF CLINTON COUNTY MICHIGAN.

Scale 3/4 of 1 inch to 1 Mile

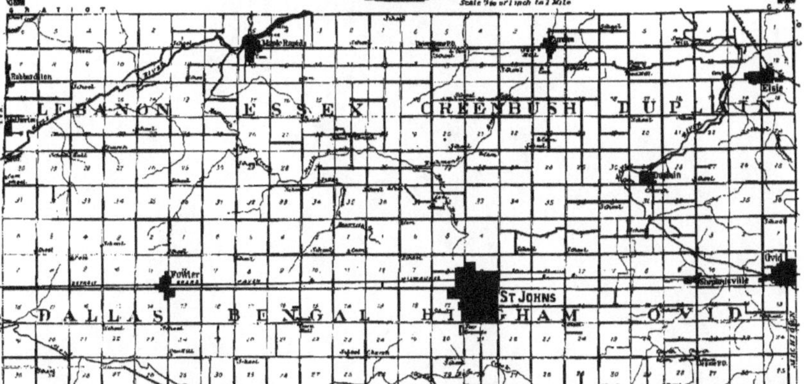

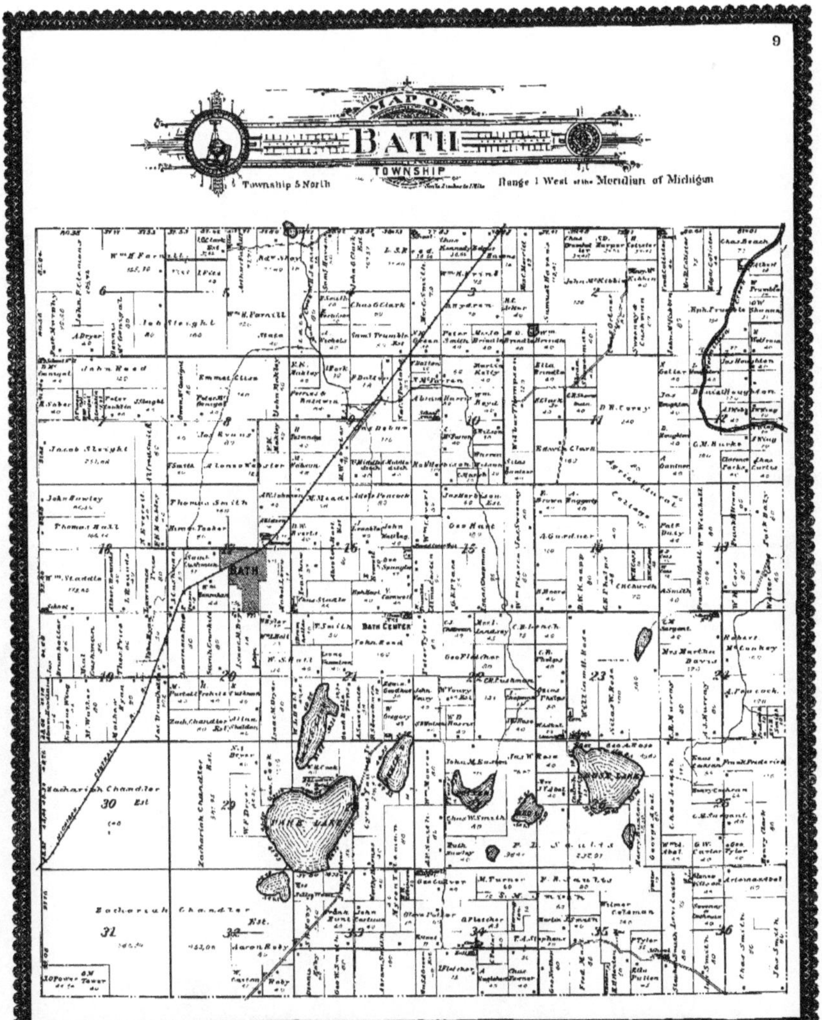
MAP OF
BATH
TOWNSHIP
Township 5 North
Range 1 West of the Meridian of Michigan

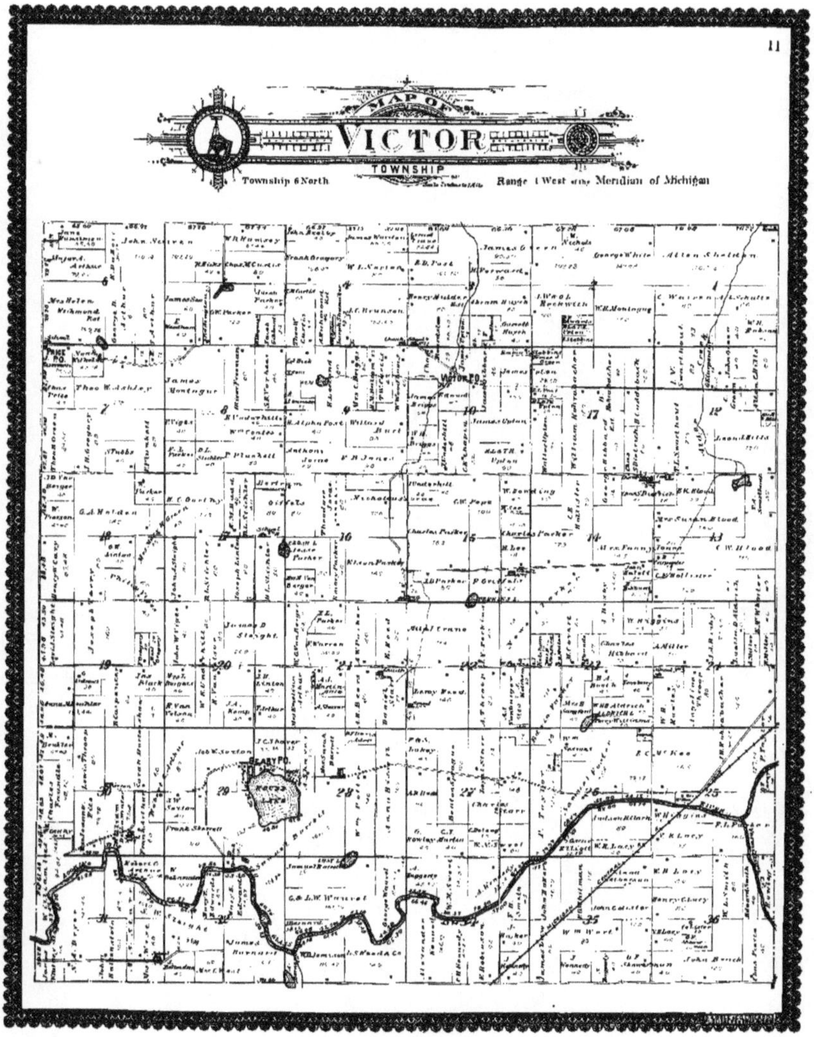
MAP OF
VICTOR
TOWNSHIP
Township 6 North
Range 1 West
Meridian of Michigan

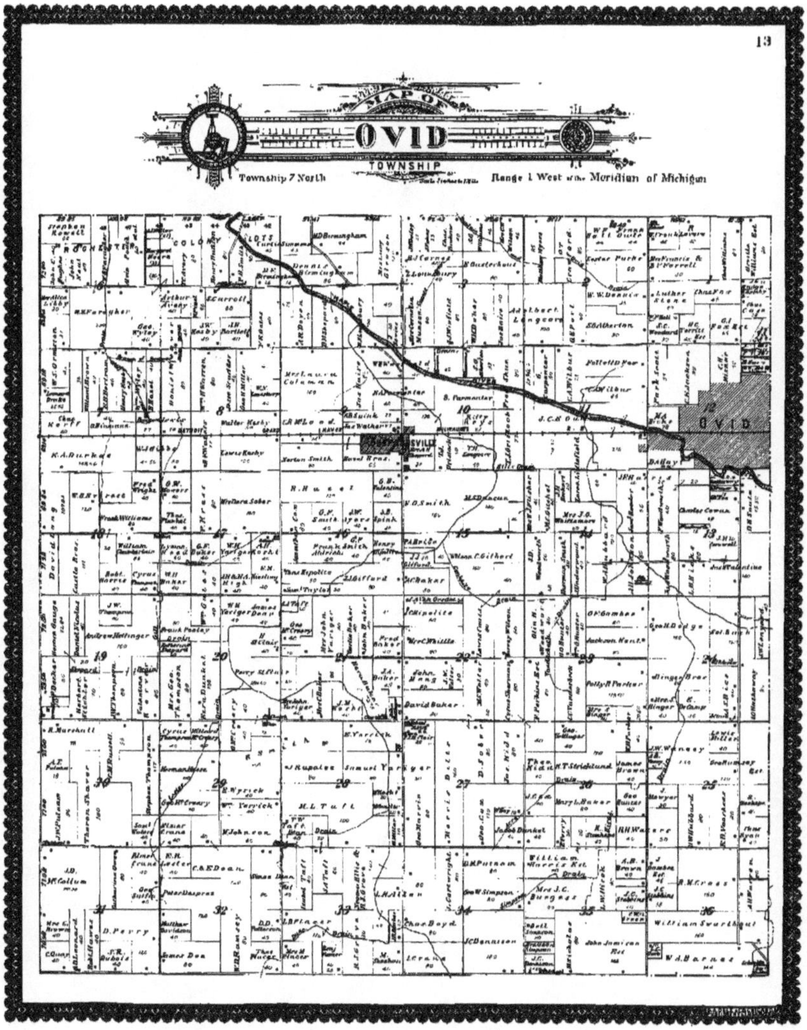
MAP OF
OVID
TOWNSHIP
Township 7 North
Range 1 West of the Meridian of Michigan
OVID

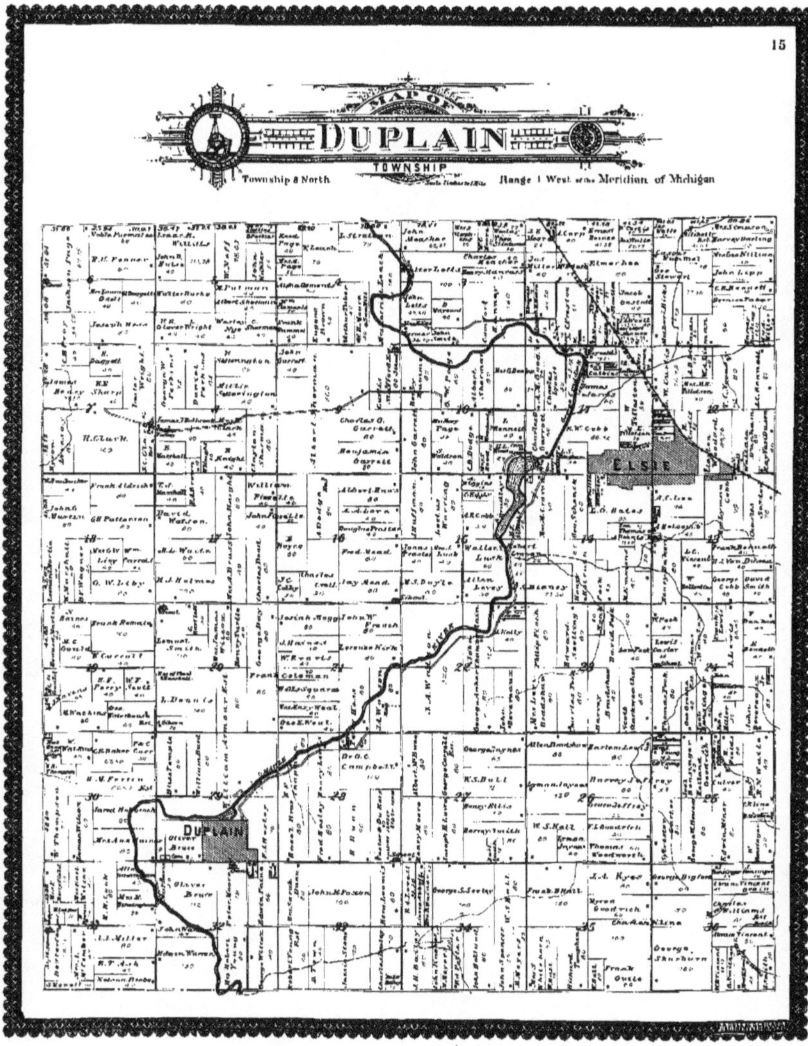
MAP OF
DUPLAIN
TOWNSHIP
Township 8 North
Range 1 West of the Meridian of Michigan
ELSIE
DUPLAIN

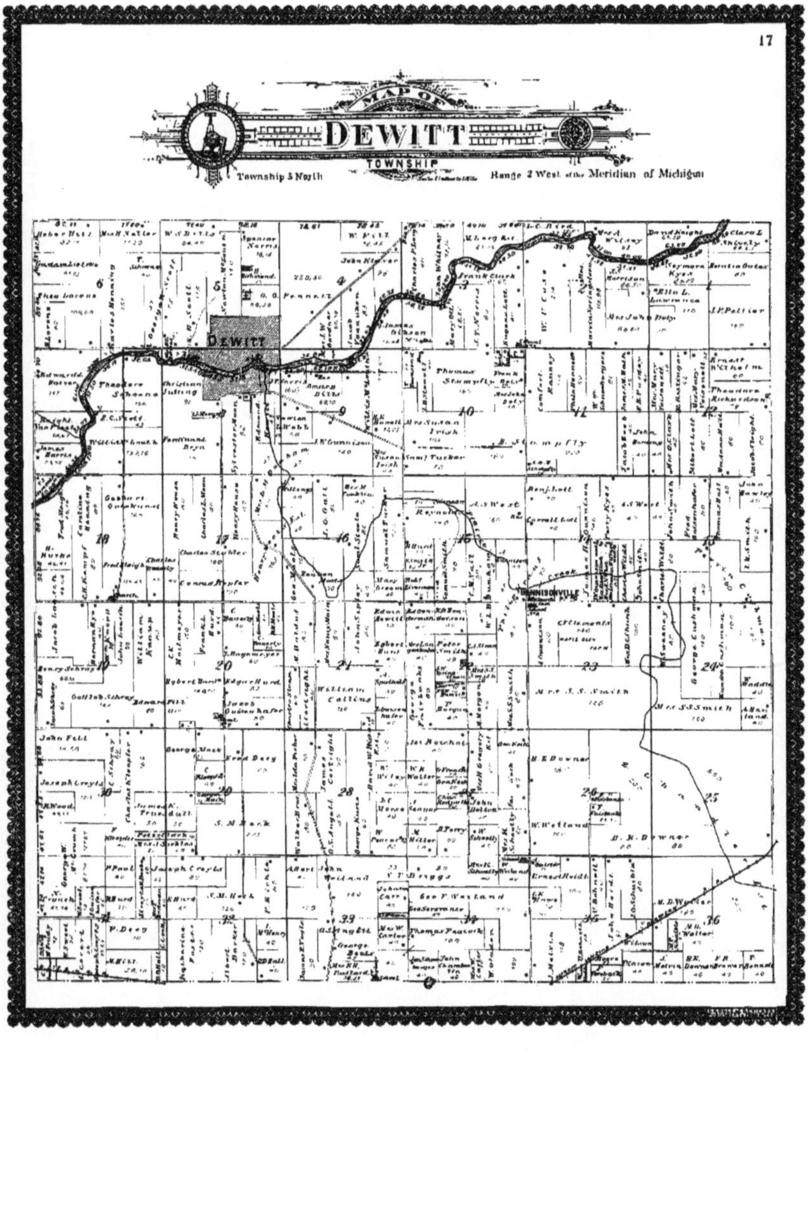
MAP OF
DEWITT
TOWNSHIP
Township 5 North
Range 2 West of the Meridian of Michigan
DEWITT
GUNNISONVILLE

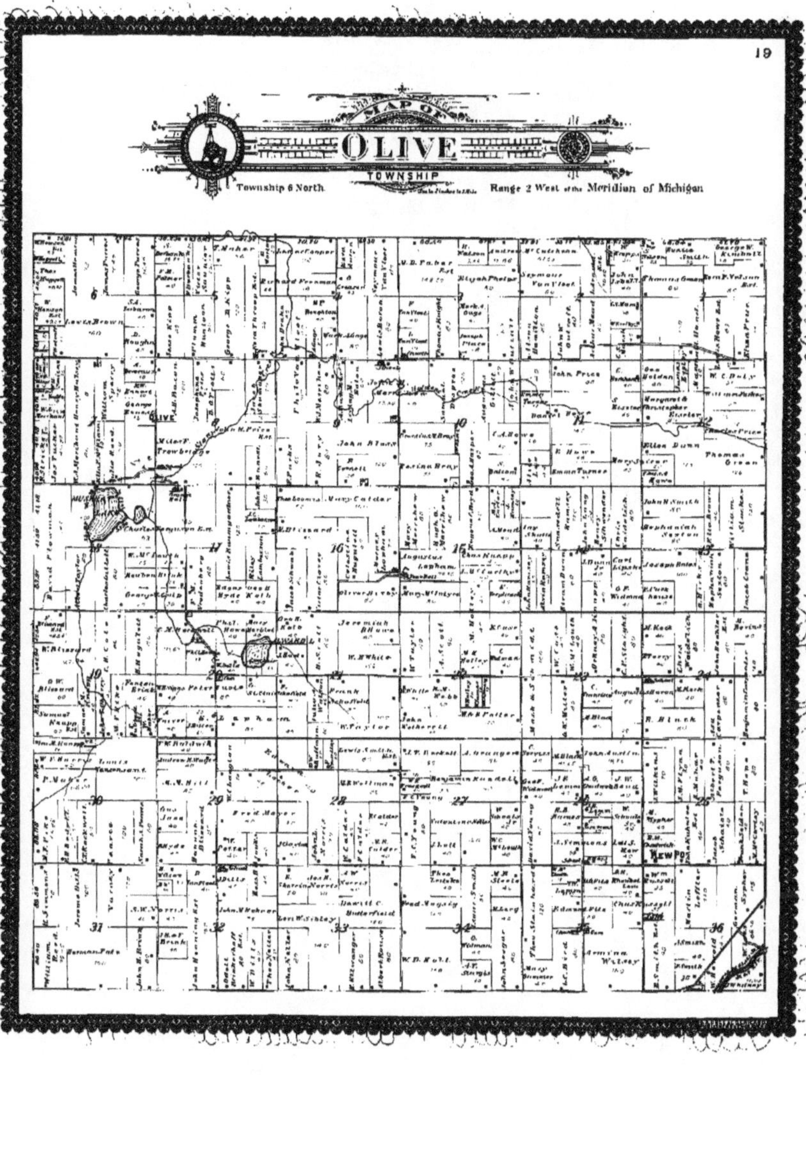
MAP OF
OLIVE
TOWNSHIP
Township 6 North.
Range 2 West of the Meridian of Michigan

MAP OF BINGHAM TOWNSHIP

Township 7 North. Range 2 West of the Meridian of Michigan

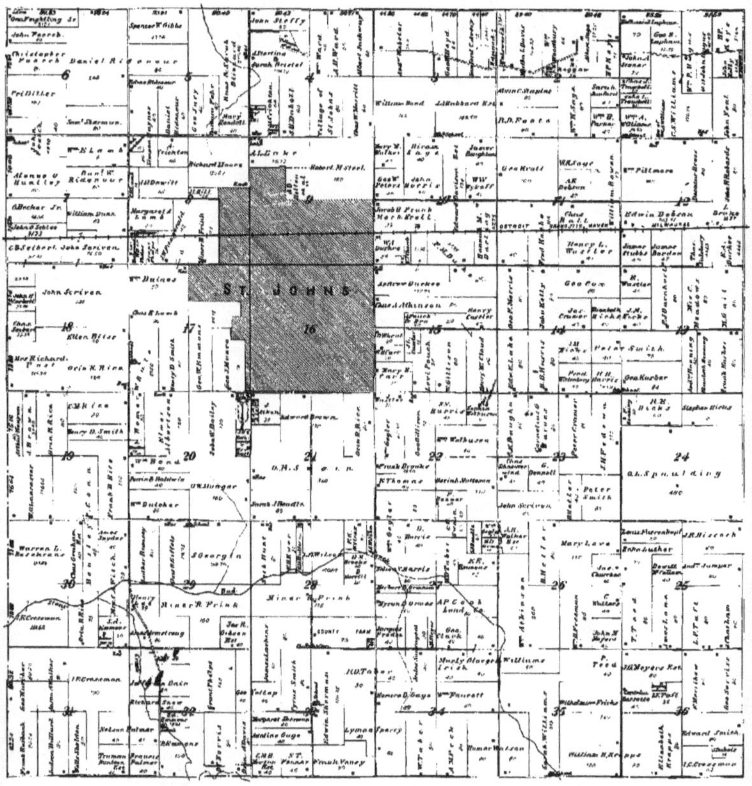

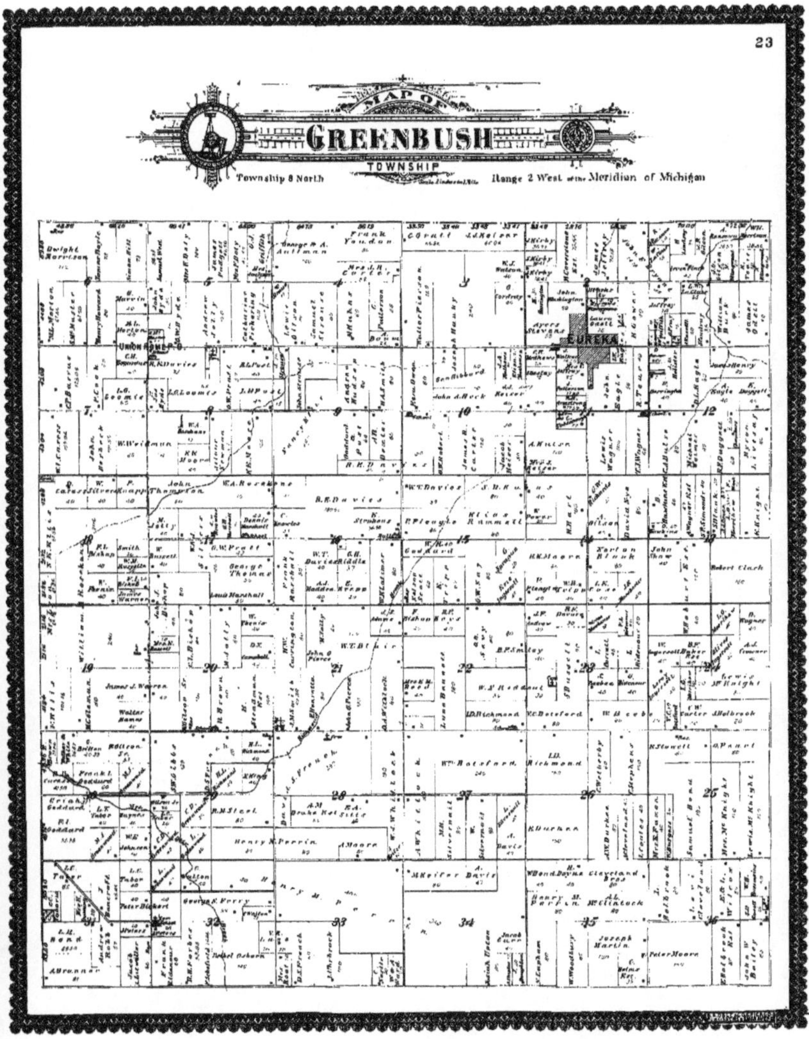
MAP OF
GREENBUSH
TOWNSHIP
Township 8 North
Range 2 West of the Meridian of Michigan
EUREKA
UNION HOME P.O.

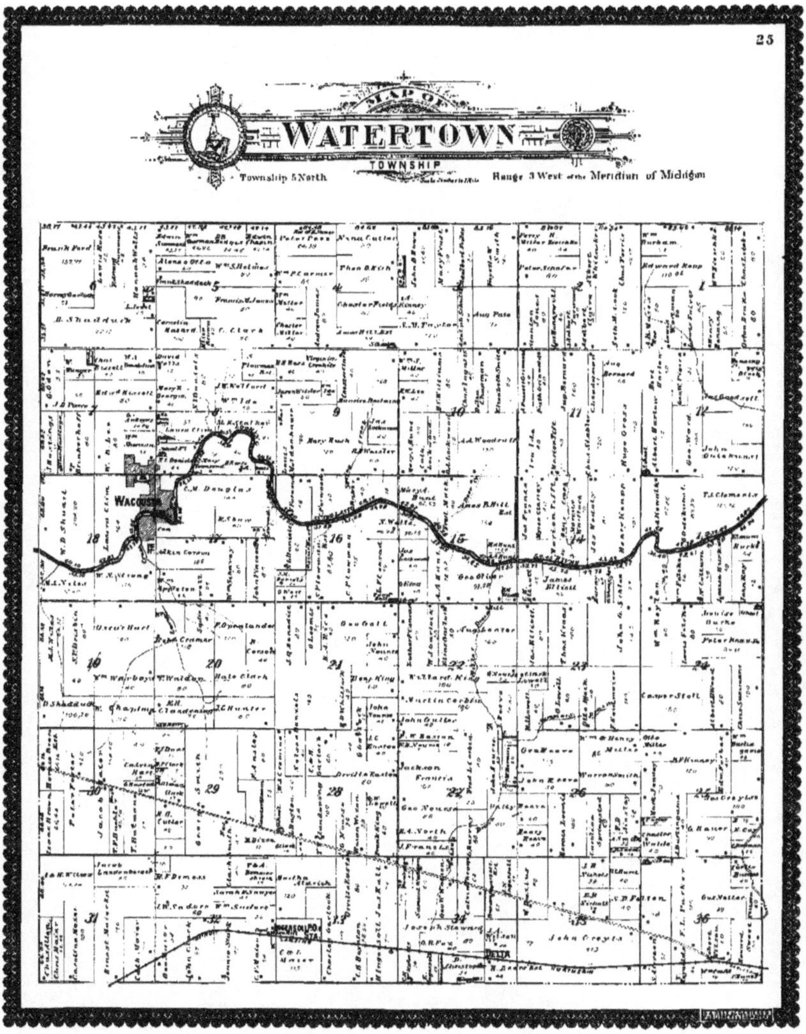
MAP OF
WATERTOWN
TOWNSHIP
Township 5 North
Range 3 West of the Meridian of Michigan
WACOUSTA
DELTA

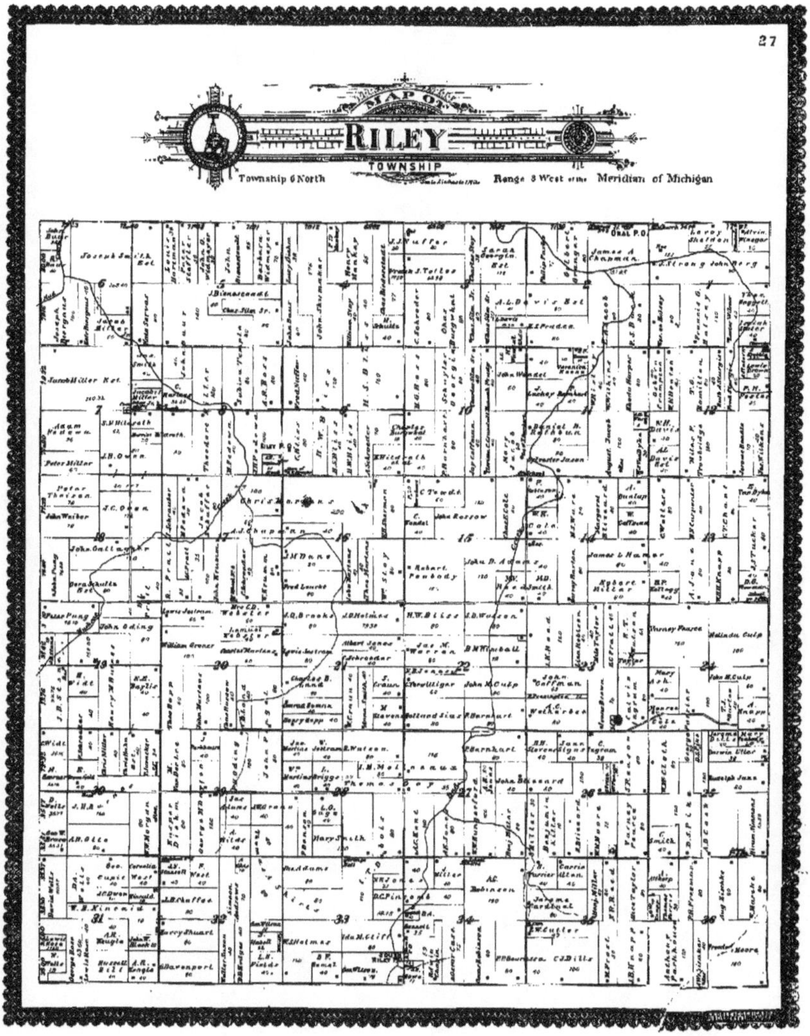
MAP OF
RILEY
TOWNSHIP
Township 6 North
Range 3 West of the Meridian of Michigan

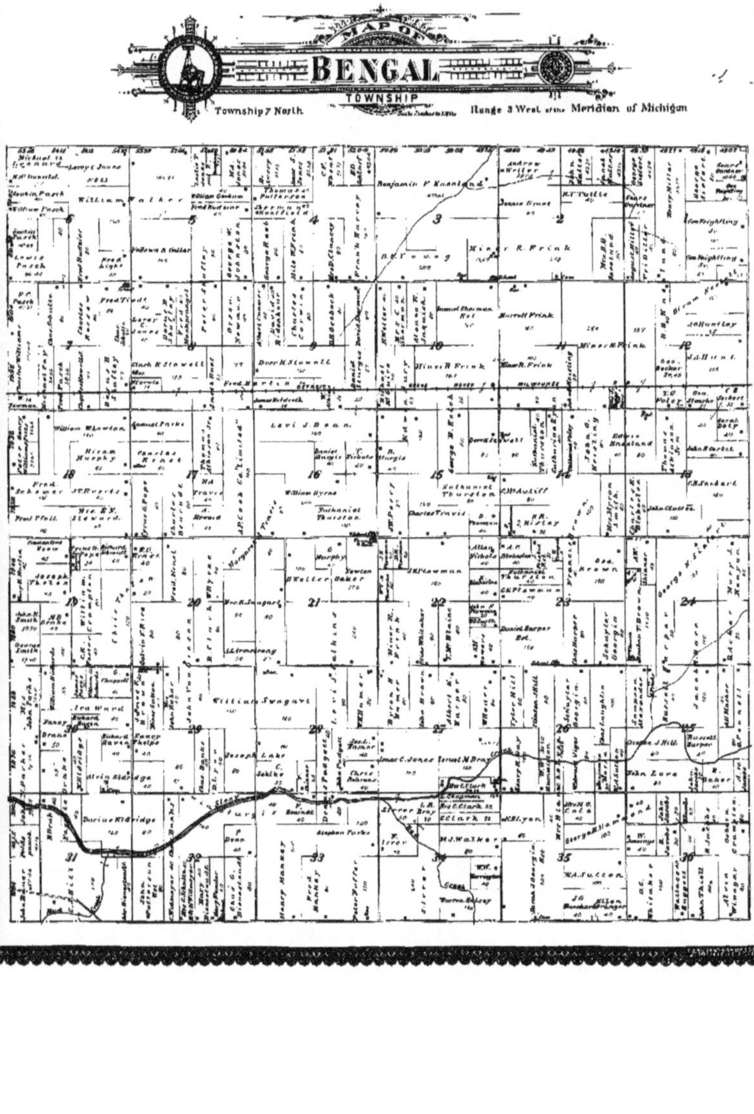
MAP OF
BENGAL
TOWNSHIP
Township 7 North.
Range 3 West of the Meridian of Michigan

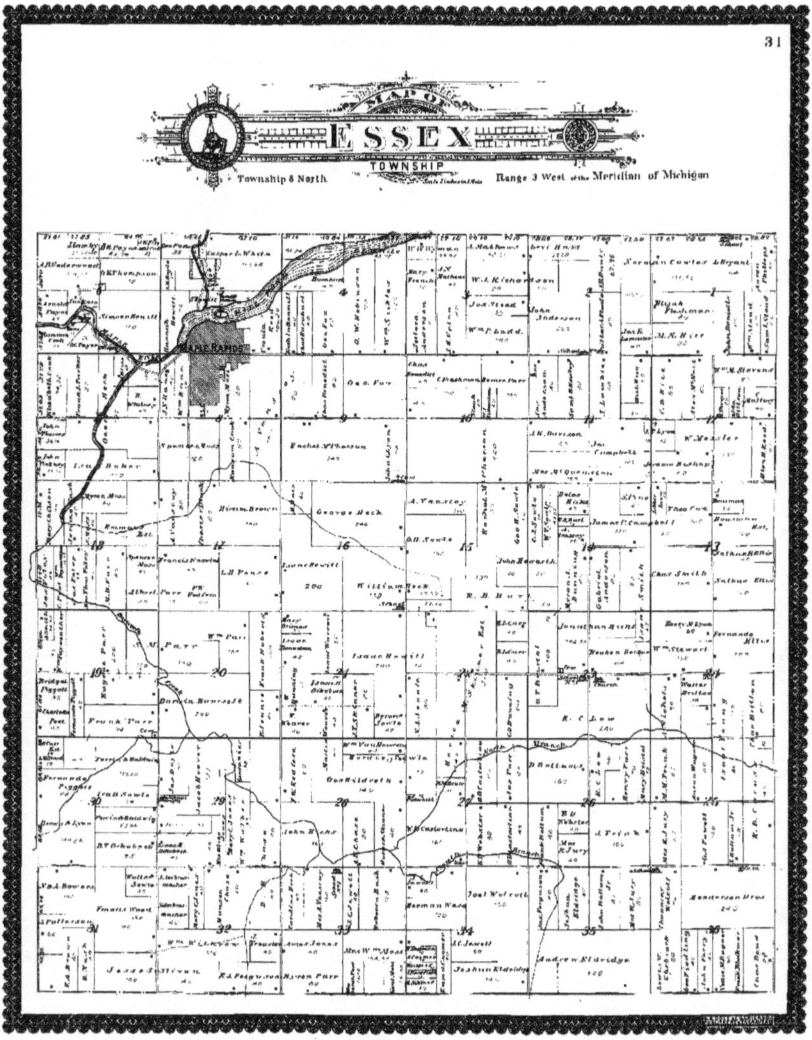
MAP OF
ESSEX
TOWNSHIP
Township 8 North
Range 3 West of the Meridian of Michigan
MAPLE RAPIDS

MAP OF EAGLE TOWNSHIP

Township 5 North Range 4 West of the Meridian of Michigan

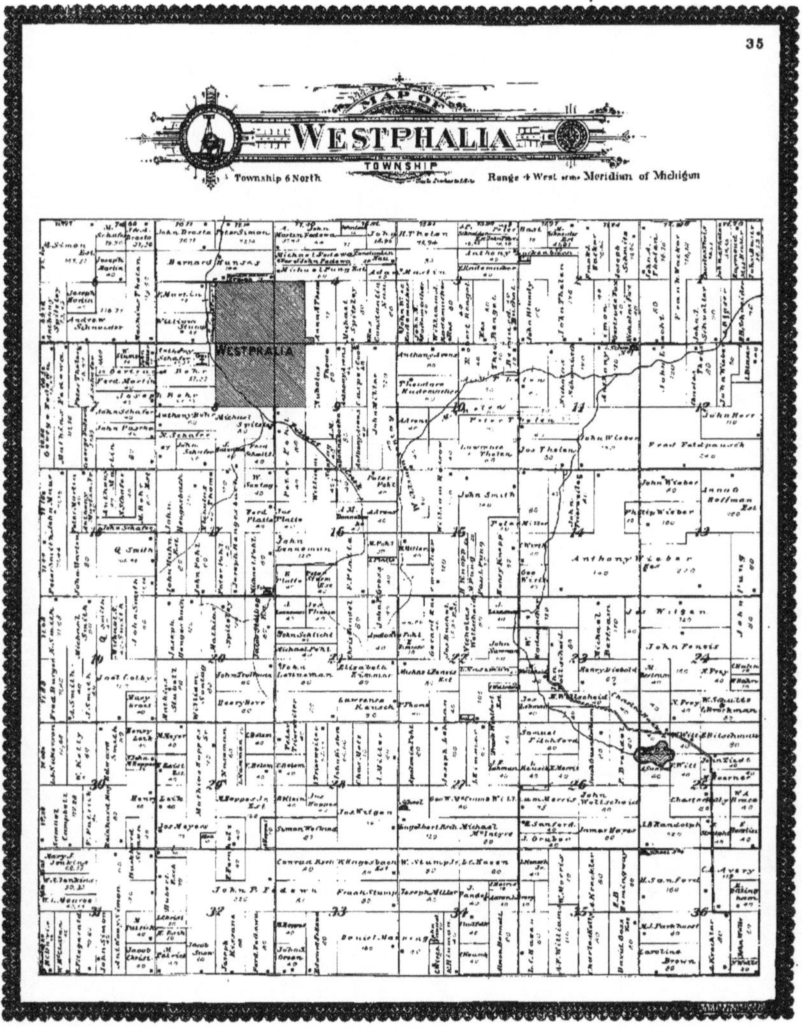
MAP OF
WESTPHALIA
TOWNSHIP
Township 6 North
Range 4 West of the Meridian of Michigan
WESTPHALIA
John Smith
Anthony Wieber
John P. Fedewa

MAP OF DALLAS TOWNSHIP

Township 7 North. Range 4 West of the Meridian of Michigan

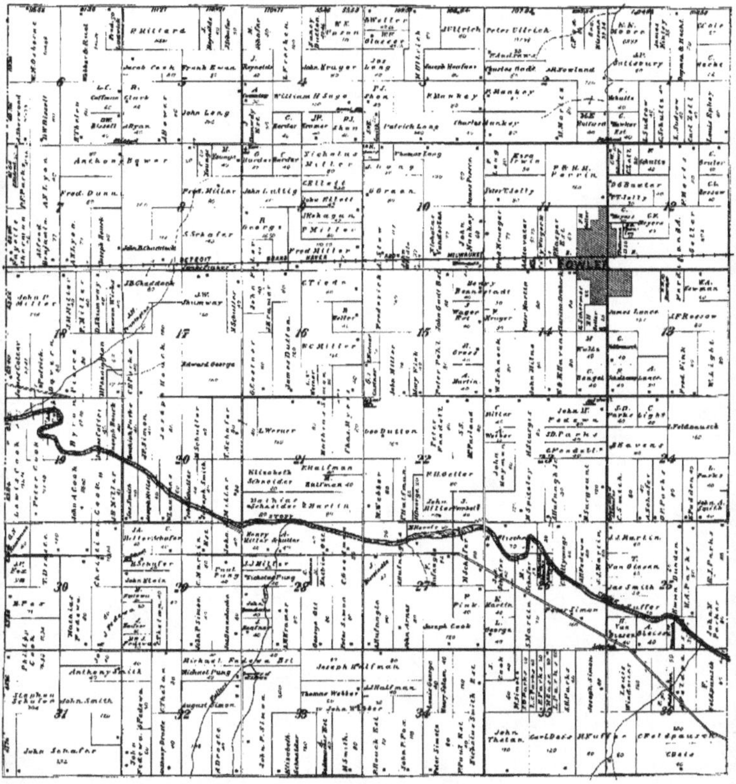

MAP OF
LEBANON
TOWNSHIP
Township 8 North
Range 4 West of the Meridian of Michigan
HUBBARDSTON
MATHERTON

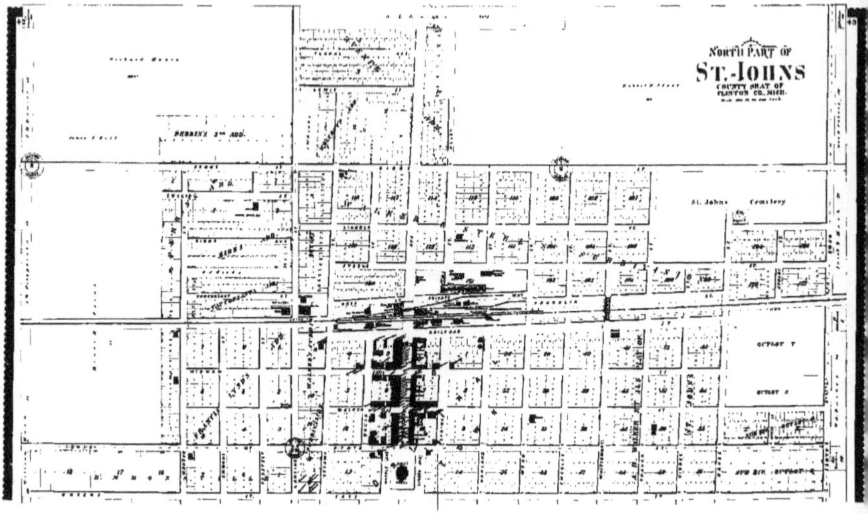
NORTH PART OF
ST. JOHNS
COUNTY SEAT OF
CLINTON CO., MICH.
St. Johns Cemetery
PERRIN'S 2ND ADD.
OUTLOT Y
OUTLOT X
RAILROAD

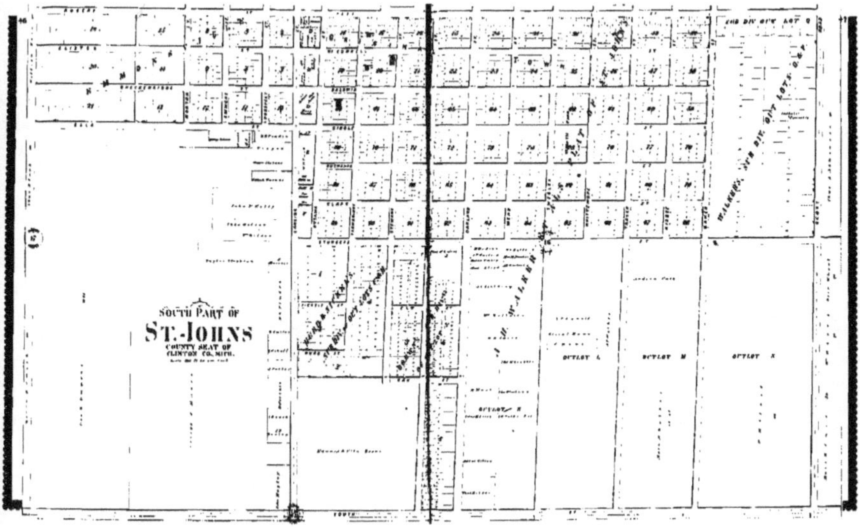
SOUTH PART OF
ST.-JOHNS
COUNTY SEAT OF
CLINTON CO., MICH.
OUTLOT L
OUTLOT M
OUTLOT N
SUB DIV OUT LOT Q
WALKERS SUB DIV. OUT LOTS O&P

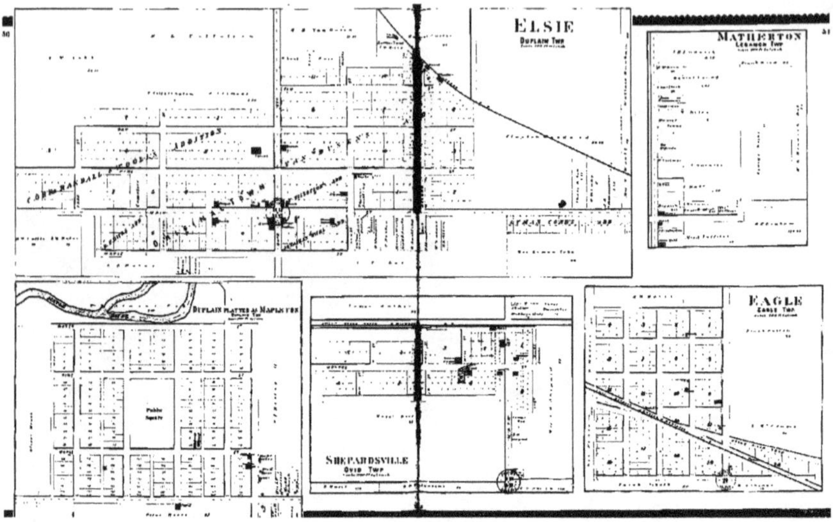
50
51
ELSIE
DUPLAIN TWP
MATHERTON
LEBANON TWP
DUPLAIN PLATTED AS MAPLETON
Public Square
SHEPARDSVILLE
OVID TWP
EAGLE
EAGLE TWP

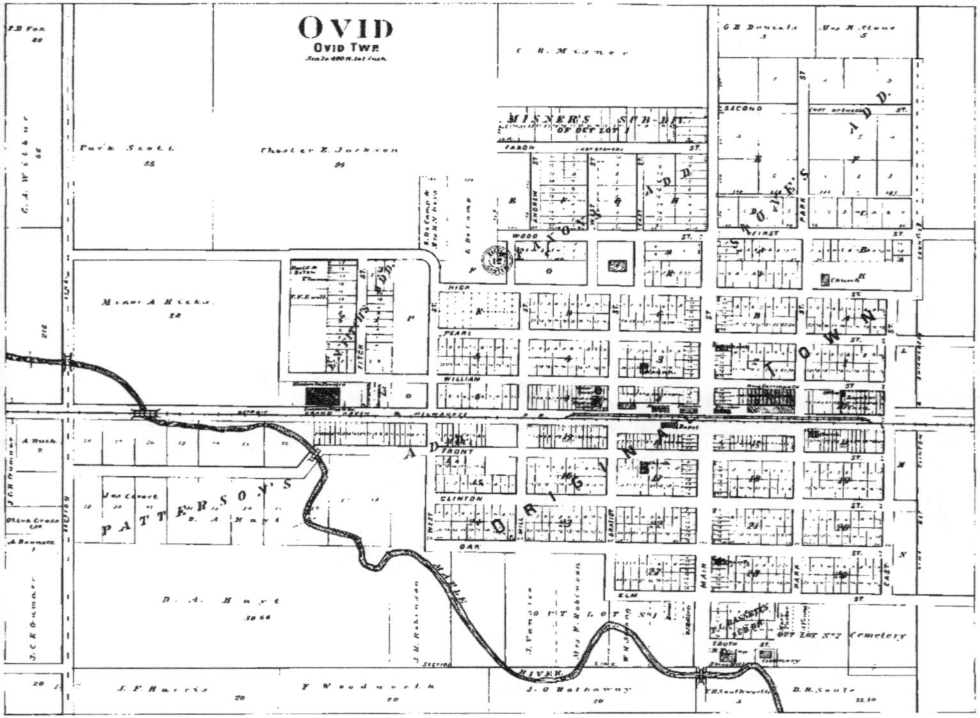
OVID
OVID TWP.
C. H. Misner
MISNER'S SUB-DIV. OF OUT LOT 1
Park Scott 55
Chester E. Jackson 99
G. A. Wilbur 66
Minor A. Hicks 28
G. B. Daniels 3
Mrs H. Stone 5
SECOND
PARK
FIRST
WOOD ST.
HIGH
PEARL
WILLIAM
FRONT
CLINTON
OAK
ELM
MAIN
EAST
Church
Depot
GRAND HAVEN & MILWAUKEE R. R.
ORIGINAL TOWN
PATTERSON'S
D. A. Hart
MAPLE RIVER
J. O. Hathaway 20
J. F. Harris 20
F. Woodworth 70
D. H. Saule
OUT LOT No 2
Cemetery
OUT LOT No 1
Mrs R. Robinson
J. H. Robinson
A. Huck
J. C. R. Gummer
A. Bennett
F. B. Fox 80

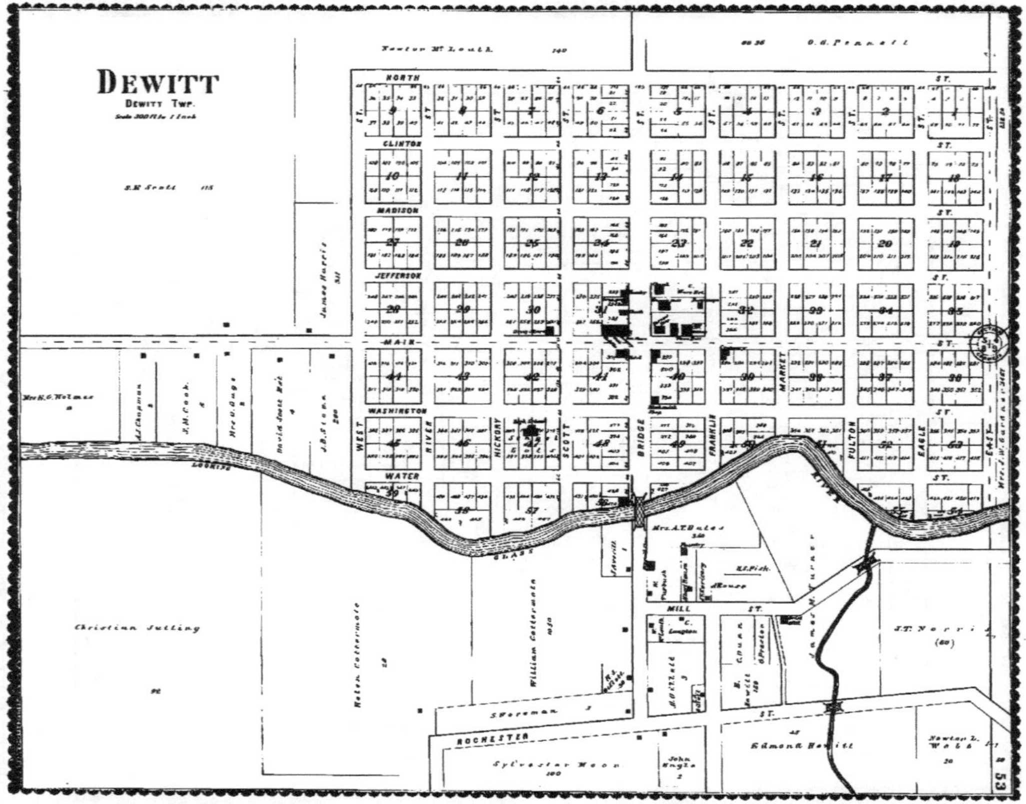
DEWITT
DEWITT TWP.
NORTH
CLINTON
MADISON
JEFFERSON
MAIN
WASHINGTON
WATER
WEST
RIVER
HICKORY
SCOTT
BRIDGE
FRANKLIN
MARKET
FULTON
EAGLE
EAST
ST.
MILL
ROCHESTER
LOOKING
GLASS
RIVER
Newton McLouth
O. G. Pennell
S. H. Scott
James Harris
J. B. Sloan
David Scott Est.
Mrs O. Gage
J. M. Cook
A. J. Chapman
Mrs H. G. Holmes
Christian Sutling
Helen Cattermole
William Cattermole
S. Foreman
Sylvester Moon
John Engle
Mrs. A. T. Bates
E. S. Pick
J. Rouse
James M. Turner
J. T. Norris
(60)
Edmond Hewitt
Newton L. Webb
53

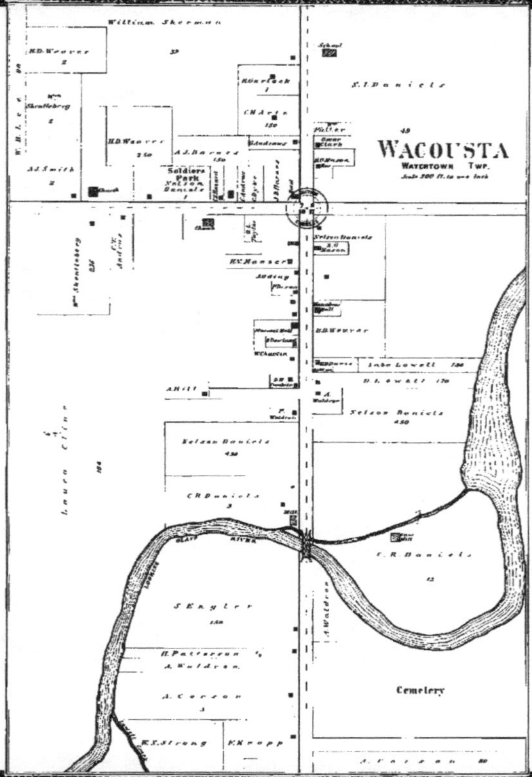
WACOUSTA
WATERTOWN TWP.
Scale 300 ft. to one inch
William Sherman
S.I. Daniels
Soldiers Park
School
Church
Nelson Daniels
C.R. Daniels
Cemetery
Looking Glass River

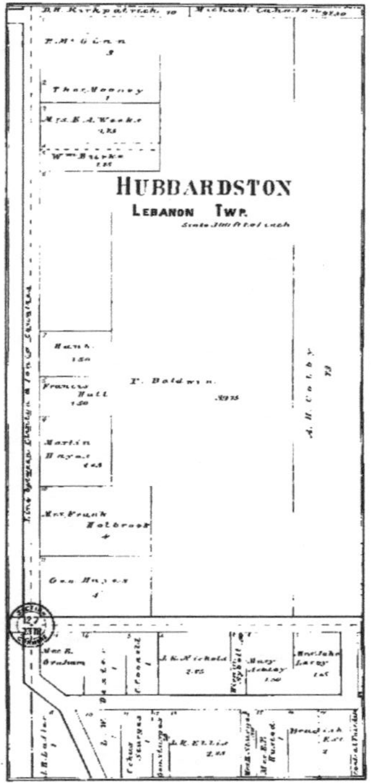
HUBBARDSTON
LEBANON TWP.
Scale 300 ft. to 1 inch
T. Baldwin
A. H. Colby
Geo Hayes

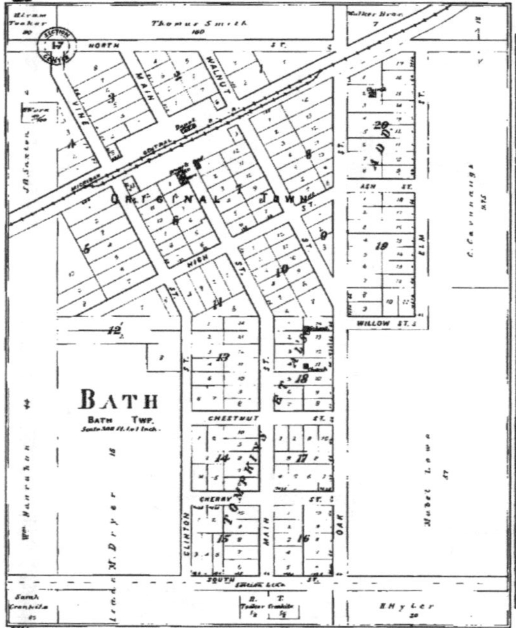
BATH
BATH TWP.
ORIGINAL TOWN
NORTH ST.
MAIN
WALNUT
VINE
HIGH ST.
CHESTNUT ST.
CHERRY
CLINTON
OAK
SOUTH ST.
ASH ST.
ELM
WILLOW ST.
Thomas Smith
Hiram Tooker
J.B. Saxton
Wm Hanrahan
M. Dryer
Mabel Lowe
C. Cavanaugh
E Hyler
Walker Bros.

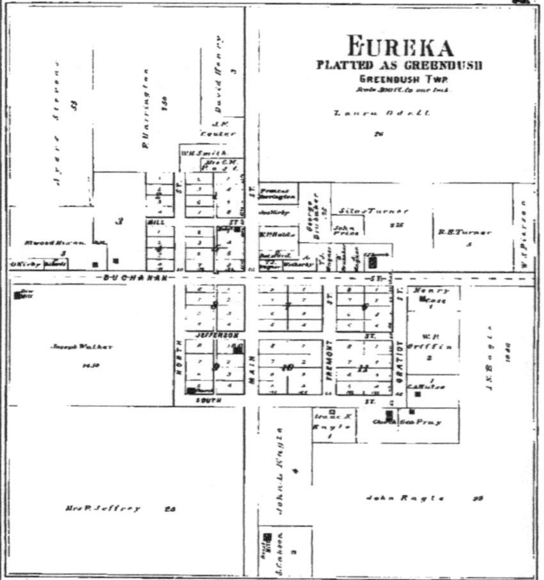
EUREKA
PLATTED AS GREENBUSH
GREENBUSH TWP.
BUCHANAN ST.
JEFFERSON
NORTH
MAIN
FREMONT
GRATIOT
SOUTH
MILL
Laura Odell
Ayers Stevens
P. Harrington
David Henry
J.K. Tooker
W.H. Smith
Silas Turner
R.B. Turner
W.T. Pierson
John Price
Joseph Walker
Mrs P. Jeffrey
John L. Kagle
John Kagle
J.N. Kagle
W.P. Griffin
Henry Case
Isaac K. Kagle
Geo Pray
J. Cahoon

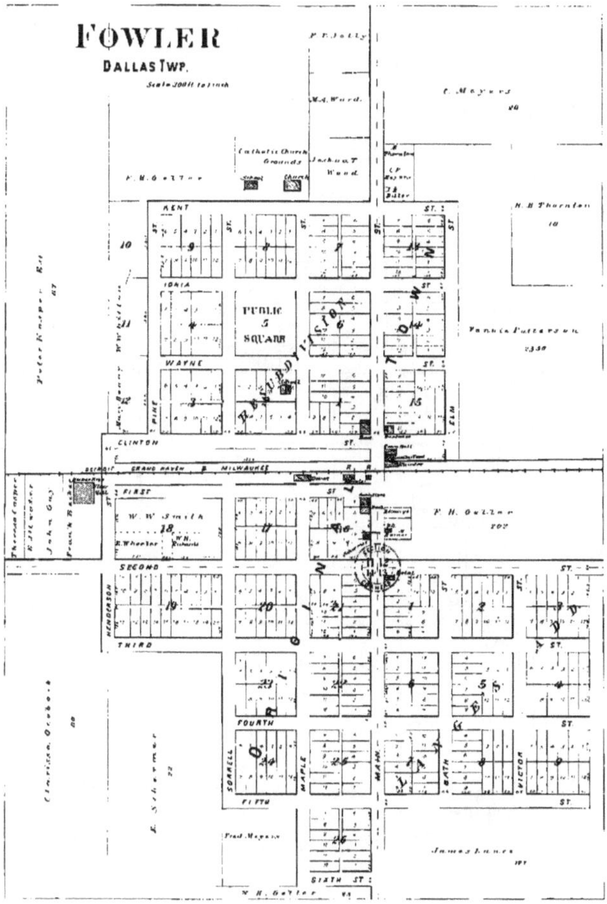
FOWLER
DALLAS TWP.
KENT
IONIA
WAYNE
PUBLIC SQUARE
CLINTON ST.
DETROIT GRAND HAVEN & MILWAUKEE R. R.
FIRST
SECOND
THIRD
FOURTH
FIFTH ST.
SIXTH ST.
HENDERSON
PINE
ELM
SORRELL
MAPLE
MAIN
BATH
VICTOR
W. W. Smith
H. B. Thornton
Fannie Patterson
F. H. Geller

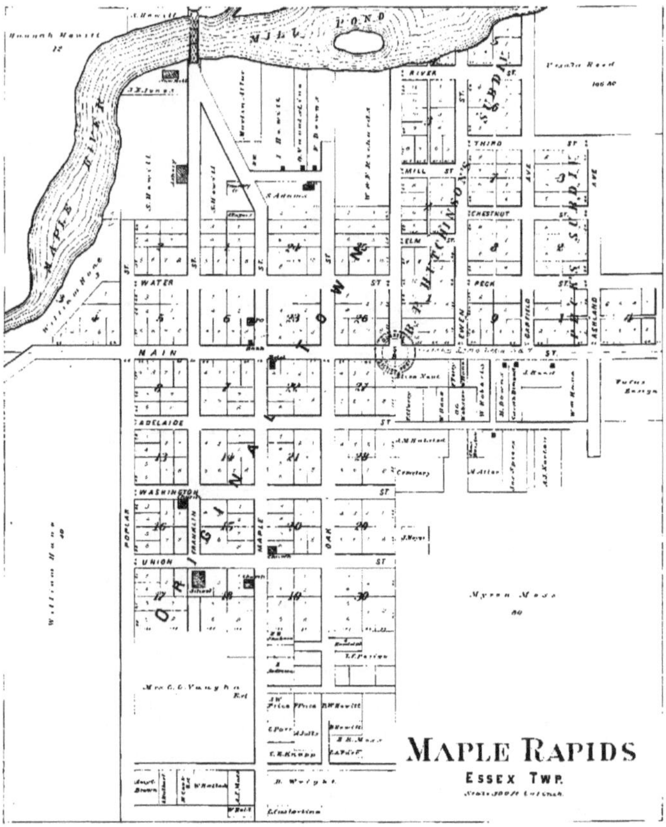
MILL POND
MAPLE RIVER
RIVER ST.
THIRD ST.
MILL ST.
CHESTNUT
ELM
WATER ST.
MAIN ST.
ADELAIDE ST.
WASHINGTON ST.
UNION ST.
PECK
EWEN
GARFIELD
ASHLAND
AVE
POPLAR
FRANKLIN
MAPLE
OAK
ORIGINAL TOWN
Cemetery
School
Myron Moss 80
MAPLE RAPIDS
ESSEX TWP.

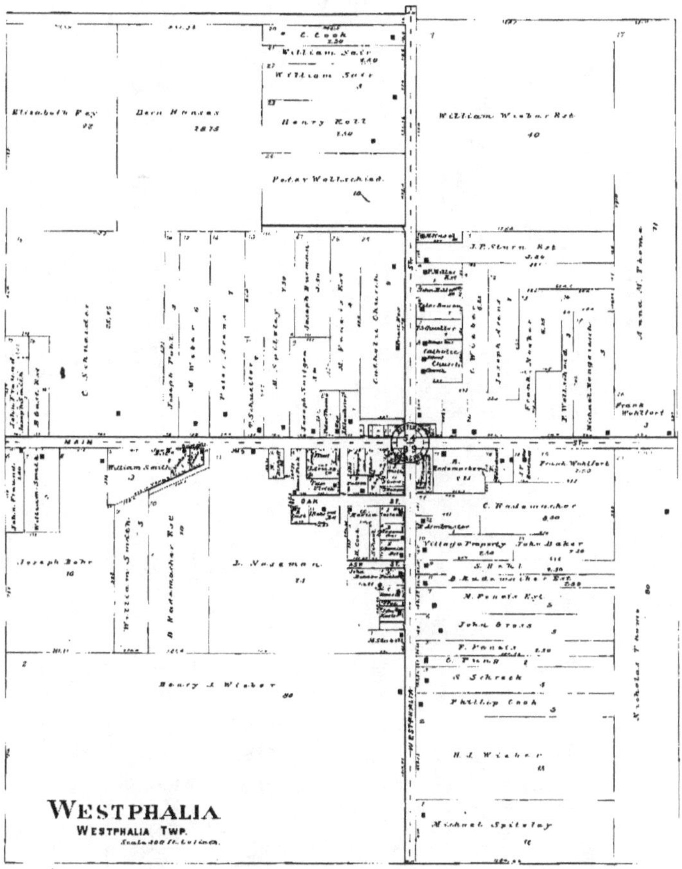
WESTPHALIA
WESTPHALIA TWP.
MAIN
OAK
Elizabeth Fay
C. Cook
William Sair
Henry Kott
Peter Wollschied
William Wieber Est.
J. P. Sturn Est.
Anna M. Thome
C. Schneider
Joseph Fohl
M. Wieber
Peter Arens
M. Spitzley
Catholic Church
Joseph Arens
Frank Wohlfort
William Smith
C. Rademacher
Village Property
John Baker
M. Fenets Est.
John Gross
S. Schrock
Phillip Cook
Joseph Bohr
Henry J. Wieber
H. J. Wieber
Michael Spitzley
Nicholas Thome

MICHIGAN.

MAP OF
MICHIGAN
Published by
GEO. A. OGLE & CO.
LAKE HURON
LAKE MICHIGAN

EXPLANATION TO INDEX

County Seats in full face type

• Money Order Post Office

X Less than 100 population

Towns in Roman Type Post Offices

Towns in *Italic* Type not Post Offices

[illegible]

[illegible]

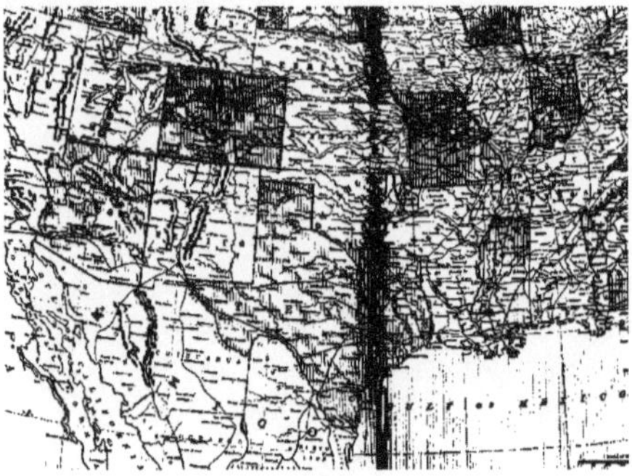

Mrs. Frank Quenken.
Bounty of
[illegible]

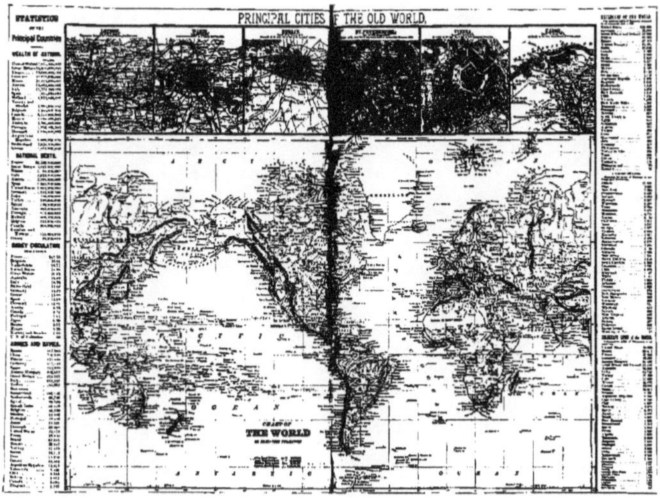
PRINCIPAL CITIES F THE OLD WORLD.
STATISTICS
Principal Countries
THE WORLD

REFERENCE DIRECTORY

OF

CLINTON ⊛ COUNTY, ⊛ MICHIGAN.

EXPLANATION.—The date following a name indicates the length of time the party has been a resident of the county. The abbreviations are as follows: S. for Section; T. for Township; and P. O., for Post-office address. When no Section Number, nor Township is given, it will be understood that the party resides within the limits of the village or city named, and, in such cases, the post-office address is the same as the place of residence, unless otherwise stated.

Abel, Bradley, Farming, S. 15, T. Duplain, P. O. Elsie, 1871.
Adams, John D., Farming & Stock, S. 14, T. Riley, P. O. Oral, 1891.
Albertson, Elmer, Farming, S. 20, T. Bingham, P. O. St. Johns, 1862.
Aldrich, W. H. H., Farming & Stock, S. 23, T. Victor, P. O. Laingsburg, 1843.
Aldridge, Wm., Hardware, Maple Rapids, 1858.
Alexander, M. S., Farming, S. 1, T. Eagle, P. O. Wacousta, 1837.
Allen, L. H., Farming & Stock, S. 34, T. Ovid, P. O. Shepardsville, 1809.
Allison, C. S., Watches, Clocks & Jewelry, St. Johns, 1880.
Anderson, James, Justice of the Peace & Farming, S. 11, T. Essex, P. O. Maple Rapids, 1865.
Anderson, G., Farming, S. 14, T. Essex, P. O. Maple Rapids, 1854.
Arens, Anthony, Farming, S. 9, T. Westphalia, P. O. Westphalia, 1841.
Armstrong, A. S., Farming, S. 21, T. Bengal, P. O. St. Johns, 1846.
Atkinson, T., Farming, S. 13, T. Bengal, P. O. St. Johns, 1861.
Atkinson, Wm., Farming, S. 26, T. Bingham, P. O. St. Johns, 1862.
Avery, Arthur, Farming, S. 5, T. Ovid, P. O. Duplain, 1864.
Avery, S. C., Farming, S. 3, T. Bingham, P. O. St. Johns, 1844.
Avery, T. C., Farming, S. 5, T. Ovid, P. O. Duplain, 1840.

Baines, Wm., Farming & Stock, S. 17, T. Bingham, P. O. St. Johns, 1880.
Bailey, J. H., Farming, S. 34, T. Duplain, P. O. Shepardsville, 1867.
Baker, N., Farming, S. 21, T. Bengal, P. O. St. Johns, 1867.
Baker, J. A., Farming, S. 21, T. Ovid, P. O. Jessie, 1858.
Baker, David, Farming, S. 22, T. Ovid, P. O. Jessie, 1866.
Baker, W. H., Farming, S. 3, T. Ovid, P. O. Shepardsville, 1855.
Baker, A. C., Farming, S. 15, T. Ovid, P. O. Shepardsville, 1860.
Baldwin, A. J. (Perrine & Baldwin), Law, Loans & Real Estate, St. Johns, 1867.
Baldwin, Levi W., Hardware Merchant & Twp. Supervisor, Fowler, 1867.
Ball, C. E. (Fowler & Ball), Dealer in General Hardware, St. Johns, 1876.
Bamber, Wm., Farming, S. 1, T. Lebanon, P. O. Maple Rapids, 1862.
Bancroft, J., Farming, S. 31, T. Greenbush, P. O. St. Johns, 1865.
Barr, J. E., Farming, S. 25, T. Bengal, P. O. St. Johns, 1855.
Barnes, C. G., Farming, S. 23, T. Bingham, P. O. St. Johns, 1867.
Barnes, N. B., Farming, S. 26, T. Olive, P. O. Rew, 1856.
Barnhart, X., Farming & Stock, S. 10, T. Riley, P. O. Oral, 1872.
Bates, J. B., Farming & Stock, S. 21, T. Eagle, P. O. Eagle, 1862.
Bates, L. G., General Merchant, Farmer & Supervisor, Elsie, 1866.
Baughn, C. E., Farming, S. 23, T. Bingham, P. O. St. Johns, 1855.
Beadle, J. A., Farming, S. 21, T. Bingham, P. O. St. Johns, 1861.
Bebee, W., Farming, S. 23, T. Greenbush, P. O. St. Johns, 1866.
Beck, J. A., Farming, S. 10, T. Greenbush, P. O. St. Johns, 1871.
Beckwith, J. W., Twp. Supervisor & Farmer, S. 2, T. Victor, P. O. Victor, 1862.
Bement, Frank A., Editor & Propr. Ovid Register-Union, Ovid, 1870.
Bengel, Theo., Farming & Stock, S. 3, T. Westphalia, P. O. Westphalia, 1865.
Bennett, L. L., Farming, S. 10, T. Duplain, P. O. Elsie, 1854.
Bennett, A. C., Retired, S. 12, T. Duplain, P. O. Elsie, 1855.
Benay, C. E., Bazaar, St. Johns, 1895.
Bensinger, Joel, Hard & Soft Wood Lumber, White Oak Bill Stuff Specialty, S. 36, T. Duplain, P. O. Ovid, 1866.
Berg, John, Farming, S. 1, T. Riley, P. O. St. Johns, 1868.
Bertram, J. P., Druggist, Drugs & Medicines, Westphalia, 1850.
Birmingham, Dennis, Farming, S. 4, T. Ovid, P. O. Shepardsville, 1856.
Birmingham, M. F., Farming, S. 5, T. Ovid, P. O. Shepardsville, 1854.
Bishop, C. L., Farming & Stock, S. 20, T. Greenbush, P. O. St. Johns, 1869.
Bishop, Jerome, Farming, S. 12, T. Essex, P. O. Maple Rapids, 1850.
Blaney, J. S., Farming, S. 15, T. Duplain, P. O. Elsie, 1857.
Bliss, John, Farming, S. 9, T. Olive, P. O. St. Johns, 1864.
Bliss, Dr. Geo. E., Physician & Surgeon, Maple Rapids, 1890.
Bliss & Dane, Agricultural Implements & Hardware, Farm Machinery, Buggies, Harness, Blankets, Etc., Fowler, 1895.
Bliss, H. G. (Bliss & Dane), Hardware & Farm Machinery, Fowler.
Bliss, H. N., Farming, S. 9, T. Riley, P. O. Riley, 1849.
Bliss, H. W., Farming, S. 9, T. Riley, P. O. Riley, 1849.
Bliss, S. C., Farming, S. 9, T. Riley, P. O. Riley, 1880.
Blissard, Enos, Proprietor Dairy, S. 5, T. Bingham, P. O. St. Johns, 1853.
Blizzard, W., Farming, S. 19, T. Olive, P. O. Dewitt, 1855.
Bohr, J. N., Farming, S. 7, T. Westphalia, P. O. Westphalia, 1862.
Bond, Chas., Farming, S. 36, T. Essex, P. O. St. Johns, 1867.
Bond, Wm., Farming, S. 3, T. Bingham, P. O. St. Johns, 1876.
Booth, D. A., Farming, S. 23, T. Victor, P. O. Laingsburg, 1856.
Boron, Lewis, Farming, S. 4, T. Olive, P. O. St. Johns, 1865.
Bosroff, Peter, Farming, S. 6, T. Olive, P. O. St. Johns, 1857.
Botsford, V. C., Farming, S. 25, T. Greenbush, P. O. St. Johns, 1858.
Botsford, W. Z., Farming, S. 27, T. Greenbush, P. O. St. Johns, 1874.
Bottum, D., Farming, S. 26, T. Essex, P. O. St. Johns, 1866.
Bottum, Frank, Farming, S. 36, T. Essex, P. O. St. Johns, 1855.
Bottum, John, Jr., Farming, S. 35, T. Essex, P. O. St. Johns, 1865.
Boughton, F. P., Farming, S. 4, T. Olive, P. O. St. Johns, 1868.
Boyd, Chas., Farming, S. 34, T. Ovid, P. O. Shepardsville, 1868.
Brass, N. W., Postmaster, Duplain, 1892.
Bray, C. M., Farming, S. 10, T. Olive, P. O. St. Johns, 1861.
Bray, I. M., Farming, S. 27, T. Bengal, P. O. St. Johns, 1857.
Briggs, A., Farming, S. 4, T. Eagle, P. O. Eagle, 1840.
Briggs, J. H., Farming, S. 9, T. Eagle, P. O. Eagle, 1858.
Brindle, Wm., Farming, S. 3, T. Bath, P. O. Bath, 1850.
Brink, J. H., Farming, S. 31, T. Olive, P. O. Dewitt, 1868.
Bristol, J. S., Farming, S. 4, T. Bingham, P. O. St. Johns, 1862.
Britton, Chas., Farming, S. 24, T. Essex, P. O. St. Johns, 1865.
Brumley, Thos. Jr., Manager St. Johns Mercantile Co., St. Johns, 1867.
Brook, J. Q., Farming, S. 21, T. Riley, P. O. Riley, 1856.
Brown, E. F., Farming, Holstein Cattle, S. 25, T. Eagle, P. O. Grand Ledge, 1853.
Brown, Isaac, Retired, S. 25, T. Eagle, P. O. Grand Ledge, 1853.
Brown, Herman, Farming, S. 17, T. Essex, P. O. Maple Rapids, 1865.
Brown, A. D., Farming, S. 35, T. Ovid, P. O. Ovid, 1868.
Brown, Francis, Farming, S. 23, T. Bengal, P. O. St. Johns, 1855.
Brown, R. T., Farming, S. 24, T. Bengal, P. O. St. Johns, 1866.
Brown, John, Farming, S. 25, T. Bengal, P. O. St. Johns, 1852.
Brunson, Will H., Attorney & Counsellor at Law, St. Johns, 1858.
Brunson, J. C., Farming, S. 4, T. Victor, P. O. Victor, 1866.
Brys, Eugene C., Farming, S. 15, T. Olive, P. O. St. Johns, 1858.
Bryant, John, Farming, S. 1, T. Greenbush, P. O. Eureka, 1854.
Buchel, Bernet, Farming, S. 10, T. Westphalia, P. O. Westphalia, 1846.
Buehler, T. F., Farming, Carpenter & Builder, S. 30, T. Watertown, P. O. Grand Ledge, 1871.
Bullard, Judson, Farming, S. 31, T. Bingham, P. O. St. Johns, 1855.
Bundy, W., Merchant, Groceries and Crockery, St. Johns, 1868.
Burbank, Frank, Farming, S. 31, T. Bingham, P. O. St. Johns, 1859.
Burnes, A. J., General Merchandise, Wacousta, 1859.
Burnes, J. H., Farming & Twp. Supervisor, S. 30, T. Riley, P. O. Wacousta, 1856.
Burnes, H. M., Farming, S. 19, T. Riley, P. O. Wacousta, 1856.
Burt, R. B., Farming, S. 15, T. Essex, P. O. Maple Rapids, 1868.
Butler, Thos. C., Township Supervisor, St. Johns, 1860.
Butterfield, D. C., Farming, S. 33, T. Olive, P. O. Dewitt, 1857.
Byam, Frank E., S. 23, T. Eagle, P. O. Grand Ledge, 1867.
Byrne, W., Farming, S. 16, T. Bengal, P. O. St. Johns, 1856.

Caress, W. L., Farming, S. 7, T. Greenbush, P. O. St. Johns, 1865.
Caress, R. B., Farming, S. 23, T. Essex, P. O. St. Johns, 1864.
Carpenter, L. G., Farming & Stock, S. 21, T. Victor, P. O. Laingsburg, 1873.
Carpenter, Asa, Farming, S. 24, T. Olive, P. O. Rew, 1861.
Carpenter, Benj., Farming, S. 24, T. Olive, P. O. Geary, 1856.
Carr, P., Farming & Stock, S. 30, T. Duplain, P. O. Duplain, 1894.
Carrington, H. W., Drain Commissioner, S. 21, T. Greenbush, P. O. St. Johns, 1865.
Carter, Levi, Farming, S. 24, T. Duplain, P. O. Elsie, 1844.
Casterline, W. B., Farming, Twp. Supervisor, S. 27, T. Essex, P. O. St. Johns, 1856.
Castle, Wm. H., Law, Loans & Real Estate, St. Johns, 1874.
Catlin, W. J., Farming, S. 27, T. Lebanon, P. O. Matherton, 1848.
Clandening, T. V., Farming & Stock, S. 28, T. Watertown, P. O. Ingersoll, 1880.
Clark, L. C., Farming & Stock, S. 7, T. Duplain, P. O. Elsie, 1873.
Clark, R. S., (Clark & Hulse), Clothiers & Gents Furnishers, St. Johns, 1860.
Clark, T. H., Lunch & Billiard Rooms, St. Johns, 1855.
Clark, Chas., Farming, S. 27, T. Bengal, P. O. St. Johns, 1852.
Clark, David, Farming, S. 24, T. Eagle, P. O. Grand Ledge, 1835.
Clements, C. F., Farming & Stock, "Maple Glen Farm," S. 23, T. Dewitt, P. O. Gunnisonville, 1866.
Clemons, John W., Drain Commissioner, St. Johns, 1879.
Cleveland, Levi, Farming, S. 36, T. Greenbush, P. O. Duplain, 1866.
Cleveland, F. D., Merchant & Postmaster, Shepardsville, 1865.
Chadwick, A. G., Farming, S. 26, T. Olive, P. O. Rew, 1858.
Chadwick, L. D., Farmer, Assistant Postmaster, Rew, 1857.
Chadwick, R. M., Farmer & Postmaster, Rew, 1859.
Chapman, J. A., Farming, S. 2, T. Riley, P. O. Oral, 1852.
Chatham, J., Farming, S. 25, T. Bingham, P. O. St. Johns, 1863.
Coleman, Wilmer, Farming & Stock, S. 35, T. Bath, P. O. Haslett, 1890.
Collister, John, Farming, S. 35, T. Victor, P. O. Laingsburg, 1846.
Cook, A. B., Farming, S. 25, T. Riley, P. O. Dewitt, 1841.
Cooley, E. H., Milling, Elsie, 1866.
Corbit, Geo. S., Pubr. Clinton Independent, St. Johns, 1857.
Corwin, Chas., Farming, S. 9, T. Bengal, P. O. St. Johns, 1865.
Cox, Jonathan, Farming & Stock, S. 16, T. Ovid, P. O. Shepardsville, 1858.
Cox, Geo., Farming, S. 27, T. Ovid, P. O. Jessie, 1858.
Cox, Theo., Farming, S. 13, T. Essex, P. O. St. Johns, 1899.
Cushman, Geo., Retired, S. 24, T. Dewitt, P. O. Bath, 1833.
Craver, Peter, Farming, S. 23, T. Bingham, P. O. St. Johns, 1855.
Crumpton, Osborn, Farming, S. 36, T. Bengal, P. O. St. Johns, 1864.
Crusser, George, Farming, S. 4, T. Olive, P. O. St. Johns, 1878.
Cressman, I. F., Farmer & Surveyor, S. 31, T. Bingham, P. O. St. Johns, 1861.
Cressman, I. C., Farming, S. 36, T. Bingham, P. O. St. Johns, 1866.
Culp, John M., Farming, S. 24, T. Riley, P. O. Dewitt, 1856.
Cushman, C. R., Farming, S. 22, T. Bath, P. O. Bath, 1838.
Cushman, Perrin, Farming, S. 24, T. Dewitt, P. O. Bath, 1850.

Daball, S. B., Circuit Judge, St. Johns, 1878.
Dane, C. (Bliss & Dane), General Hardware & Implements, Fowler, 1895.
Daniels, John, Farming, S. 1, T. Essex, P. O. St. Johns, 1836.
Darling, Harvey, Farming, S. 1, T. Duplain, P. O. Elsie, 1869.
Davies, W. T., Farming, S. 15, T. Greenbush, P. O. St. Johns, 1856.
Dean, James, Farming & Stock, S. 20, T. Ovid, P. O. Shepardsville, 1866.
Dean, L. J., Farming, S. 16, T. Bengal, P. O. St. Johns, 1855.
Decker, Reuben, Farming, S. 23, T. Essex, P. O. St. Johns, 1864.
Depres, Samuel, Farming, S. 10, T. Olive, P. O. St. Johns, 1859.
Dewitt, O. P., Merchant, Groceries, Etc., St. Johns, 1864.
Dewitt Milling Co., Roller Mills, W. McLouth, Manager, Dewitt, 1866.
Dexter, R. C., Assistant Cashier St. Johns Nat'l Bank, St. Johns, 1859.
Dickerman, D., Farming, S. 2, T. Lebanon, P. O. Maple Rapids, 1858.
Dills, Omar B., Farming & Stock, S. 9, T. Dewitt, P. O. Dewitt, 1852.
Dills, W. S., Farmer & Stock Dealer, S. 5, T. Dewitt, P. O. Dewitt, 1852.
Dills, J., Farming, S. 32, T. Olive, P. O. Dewitt, 1852.
Dills, Morris, Farming & Stock, S. 27, T. Ovid, P. O. Shepardsville, 1855.
Dodge, Arthur, Farming & Stock, S. 16, T. Duplain, P. O. Elsie, 1867.
Dodge, J. M., Propr. Hotel Steel, St. Johns, 1880.
Dodge, C. H., Farming, S. 10, T. Duplain, P. O. Elsie, 1864.
Dooling, E. L., Real Estate & Collections, St. Johns, 1871.
Dooling, John C. (Lyon & Dooling), Attorney & Counsellor at Law, St. Johns, 1868.
Doty, S. P., Farming & Stock, S. 14, T. Eagle, P. O. Eagle, 1850.
Downer, B. E., Farming & Stock, S. 26, T. Dewitt, P. O. North Lansing, 1851.
Doyen, A. K., Farming, S. 4, T. Ovid, P. O. Shepardsville, 1853.
Doyle, Jas., Farming, S. 29, T. Essex, P. O. Maple Rapids, 1866.
Doyne, H., Farming, S. 35, T. Greenbush, P. O. St. Johns, 1865.
Drake, F., Farming, S. 30, T. Bengal, P. O. Fowler, 1840.
Droste, Anthony, Farming, S. 32, T. Dallas, P. O. Westphalia, 1859.
Dryer, Allen, Farming & Stock, S. 6, T. Bath, P. O. Bath, 1850.
Duncan, M. N., Justice of the Peace, St. Johns, 1867.
Dunkel, Ezra, Farming, S. 20, T. Ovid, P. O. Jessie, 1861.
Dunkel, Jacob, Farming, S. 27, T. Ovid, P. O. Ovid, 1854.
Dunning, C. A., Farming, S. 22, T. Essex, P. O. St. Johns, 1865.
Durham, John, Farming & Stock, S. 35, T. Eagle, P. O. Grand Ledge, 1874.

Easton, J. M., Farming, S. 27, T. Bath, P. O. Bath, 1856.
Eaton, I. M., Farming, S. 9, T. Olive, P. O. St. Johns, 1856.
Eddy, Charles, County Surveyor and Twp. Clerk, Jeweler, Elsie, 1879.
Eddy, W. M., Postmaster & Merchant, Fancy & Staple Groceries, South Riley, 1850.
Eddy, Luther, Farming, S. 9, T. Eagle, P. O. Eagle, 1845.
Eddy, Norris, Farming, S. 9, T. Eagle, P. O. Eagle, 1844.
Edwards, E. F., Postmaster, Victor, 1882.
Feeley, Henry, Milling, Maple Rapids, 1890.
Eldridge, Alvin, Farming, S. 30, T. Bengal, P. O. Fowler, 1840.
Eldridge, Darius, Farming, S. 31, T. Bengal, P. O. Fowler, 1842.
Eldridge, Joshua, Farming, S. 34, T. Essex, P. O. St. Johns, 1859.
Eldridge, A., Farming, S. 35, T. Essex, P. O. St. Johns, 1859.
Eldridge, W., Farming, S. 30, T. Bengal, P. O. Fowler, 1866.
Ellis, Fernando, Farming, S. 24, T. Essex, P. O. St. Johns, 1860.
Ellis, Nathan, Farming, S. 13, T. Essex, P. O. St. Johns, 1860.
Emmons, E. E., Farming, S. 27, T. Bingham, P. O. St. Johns, 1855.
Emmons, G. W., Retired, St. Johns, 1852.
Emmons, P., Farming, S. 32, T. Bingham, P. O. St. Johns, 1860.
Enseal, J. W., Farming, S. 8, T. Olive, P. O. St. Johns, 1854.
Ernst, Charles, Farming & Stock, S. 17, T. Bengal, P. O. Fowler, 1872.
Everest, W. B., Farming, S. 18, T. Ovid, P. O. St. Johns, 1849.

Faragher, W. H., Farming & Stock, S. 6, T. Ovid, P. O. Shepardsville, 1852.
Farnsworth, W. P., Mason, Bath, 1872.
Fedewa, John, Farming & Stock, S. 4, T. Westphalia, 1854.
Fedewa, J. H. (Fedewa & Walbridge), Attorney at Law, St. Johns, 1849.
Fedewa, Stephen, Farming, S. 33, T. Dallas, P. O. Westphalia, 1848.
Fedewa, John, Farming, S. 31, T. Dallas, P. O. Westphalia, 1848.
Fedewa, Mathias, Farming, S. 36, T. Dallas, P. O. Westphalia, 1849.
Fedewa, Geo., Farming, S. 7, T. Westphalia, P. O. Westphalia, 1841.
Fedewa, Edward, Farming, S. 33, T. Westphalia, P. O. Westphalia, 1858.
Feightling, Geo., Jr., Farming, S. 1, T. Bengal, P. O. St. Johns, 1862.
Finch, Philip, Farming, S. 23, T. Duplain, P. O. Elsie, 1853.
Finla, Geo., Farming, S. 5, T. Eagle, P. O. Portland, 1852.
Fitzgerald, H. H., Propr. St. Johns News, St. Johns, 1871.
Fitzgerald, J. W., Cashier State Bank, St. Johns, 1866.
Fisselle, John, Farming, S. 17, T. Duplain, P. O. Duplain, 1882.
Fletcher, Geo. W., Farming, S. 22, T. Bath, P. O. Bath, 1846.
Flynn, J. C., Attorney (Patterson & Flynn), St. Johns, 1895.
Foerch, C., Farming, S. 6, T. Bingham, P. O. St. Johns, 1866.
Foerch, John, Farming, S. 6, T. Bingham, P. O. St. Johns, 1867.

Foley, James, Farming & Stock, S. 13, T. Lebanon, P. O. Maple Rapids, 1858.
Foley, T. G., Farming, S. 12, T. Bengal, P. O. St. Johns, 1867.
Forbes, R. H., Farming, S. 32, T. Greenbush, P. O. St. Johns, 1852.
Fowler, Chas., Hardware Dealer (Fowler & Bull), St. Johns, 1861.
Fox, N., Farming, S. 2, T. Westphalia, P. O. Westphalia, 1850.
Fox, F. D., Farming, S. 11, T. Ovid, P. O. Ovid, 1868.
Francisco, C., Farming, S. 23, T. Olive, P. O. Rew, 1853.
French, D. S., Secretary & Manager St. Johns Manufacturing Co., St. Johns, 1871.
Freund, John, Farming, S. 8, T. Westphalia, P. O. Westphalia, 1851.
Frink, Miner H., Farming & Stock, S. 3, T. Bengal, P. O. St. Johns, 1839.
Frink, Marrett, Farming & Stock, S. 11, T. Bengal, P. O. St. Johns, 1847.
Frink, J., Farming, S. 26, T. Essex, P. O. St. Johns, 1860.
Fritz, Peter A., Propr. P. A. Fritz Brewing Co. (Est. 1868), Westphalia, 1895.

Gage, H. B., Farming, S. 34, T. Bingham, P. O. St. Johns, 1857.
Gage, M. A., Farming, S. 4, T. Olive, P. O. St. Johns, 1854.
Garlock, Elias, Farming, S. 22, T. Watertown, P. O. Wacousta, 1847.
Garlock, J. W., Farming, S. 22, T. Watertown, P. O. Wacousta, 1847.
Gates, F. E., Farming, S. 5, T. Ovid, P. O. Shepardsville, 1860.
Gates, Wm., Farming, S. 20, T. Ovid, P. O. Shepardsville, 1857.
Gay, H. R., Farming, S. 26, T. Bengal, P. O. St. Johns, 1854.
Gegler, Wm., Farming, S. 22, T. Bingham, P. O. St. Johns, 1867.
Georgia, R., Farming, S. 29, T. Bingham, P. O. St. Johns, 1855.
Gibson, James, Farming, S. 1, T. Dewitt, P. O. Dewitt, 1872.
Giffels, Chas. B., Farming & Stock, S. 29, T. Bingham, P. O. St. Johns, 1855.
Giffels, Bartram, Farming, S. 17, T. Victor, P. O. Victor, 1854.
Gilbert, Wilson C., Farming & Stock, S. 15, T. Ovid, P. O. Ovid, 1859.
Gillison, Geo., Farming & Stock, S. 22, T. Bingham, P. O. St. Johns, 1863.
Gillson, D. M., Farming & Stock, S. 20, T. Greenbush, P. O. St. Johns, 1869.
Gillson, S. D., Farming, S. 14, T. Greenbush, P. O. Eureka, 1868.
Gilson, W. H., Farming, S. 29, T. Greenbush, P. O. St. Johns, 1878.
Goodrich, F. J., Farming, S. 26, T. Duplain, P. O. Ovid, 1866.
Grant, L., Dealer in Agricultural Implements & Machinery, St. Johns, 1881.
Green, Chas., Farming, S. 12, T. Victor, P. O. Laingsburg, 1865.
Green, John, Farming, S. 12, T. Victor, P. O. Laingsburg, 1862.
Greenwood, C. D., Farming, S. 29, T. Greenbush, P. O. St. Johns, 1857.
Griffith, G. J., Farming, S. 3, T. Greenbush, P. O. Union Home, 1866.
Gross, J. P., Farming, S. 21, T. Westphalia, P. O. Westphalia, 1844.
Groler, Frank, President State Savings Bank, Fowler, 1861.
Gunderman, E. A., Farming, S. 22, T. Dewitt, P. O. North Lansing, 1847.
Gunnison, A. G., Farming, S. 23, T. Dewitt, P. O. Gunnisonville, 1836.
Gunnison, J. H., Farming, S. 14, T. Dewitt, P. O. Gunnisonville, 1837.

Hale, H. W., Farming, S. 30, T. Bingham, P. O. St. Johns, 1847.
Hall, Francis, Staple & Fancy Groceries, Matherton, 1860.
Hall, F. B., Farming, S. 15, T. Duplain, P. O. Duplain, 1864.
Hall, W. S., Farming, S. 27, T. Duplain, P. O. Duplain, 1864.
Halsey, Warren, Farming, S. 34, T. Bengal, P. O. St. Johns, 1853.
Haire, James, Farming & Stock, S. 9, T. Ovid, P. O. Shepardsville, 1857.
Hamer, W. K., Farming, S. 28, T. Bengal, P. O. St. Johns, 1858.
Hamilton, H., Farming, S. 2, T. Olive, P. O. St. Johns, 1867.
Hammond, G. E., Farming, S. 35, T. Bengal, P. O. St. Johns, 1853.
Hand, Marshall, Farming, S. 1, T. Olive, P. O. Price, 1854.
Hankey, Joseph, Farming, S. 3, T. Greenbush, P. O. Eureka, 1865.
Hanses, Bernard, Farming, S. 5, T. Westphalia, P. O. Westphalia, 1844.
Harden, John, Farming, S. 3, T. Lebanon, P. O. Maple Rapids, 1856.
Harper, A. K., Farming, S. 27, T. Bengal, P. O. St. Johns, 1865.
Harper, G. A., Farming, S. 10, T. Olive, P. O. St. Johns, 1853.
Harper, Russell, Farming, S. 25, T. Bengal, P. O. St. Johns, 1866.
Harrison, J., Veterinary Surgeon, (Grad. Ontario Vet. Col.), St. Johns, 1893.
Haase, Herman, Farming & Stock, S. 29, T. Ovid, P. O. Ovid, 1867.
Haviland, A. C., Farming, S. 24, T. Dewitt, P. O. Bath, 1856.
Haynes, Simon, Farming, S. 5, T. Bingham, P. O. St. Johns, 1868.
Hazen, L. C., Farming, S. 35, T. Westphalia, P. O. Eagle, 1854.
Henderson, E. B., Farming, S. 22, T. Dewitt, P. O. Gunnisonville, 1852.
Hengesbach, John, Farming, S. 8, T. Westphalia, P. O. Westphalia, 1851.
Herr, Henry, Farming, S. 20, T. Westphalia, P. O. Westphalia, 1852.
Hewitt, Isaac, Banker, Maple Rapids, 1856.
Heyd, G. M., Farming, S. 3, T. Bingham, P. O. St. Johns, 1886.
Heyer, Gustave, Clerk, Westphalia.
Hicok, G. B., Farming, S. 35, T. Ovid, P. O. Ovid, 1870.
Hickok, Geo., Carpenter & Builder, Maple Rapids, 1870.
Hicks, O. E., Farming & Stock, S. 1, T. Duplain, P. O. Elsie, 1865.
Hicks, Stephen, Farming, S. 24, T. Bingham, P. O. St. Johns, 1866.
Hicks, John M., Farming, S. 13, T. Bingham, P. O. St. Johns, 1865.
Hicks, Jonathan, Farming, S. 23, T. Essex, P. O. St. Johns, 1863.
High, H. M., Attorney & Counsellor at Law, Ovid, 1882.
Hildreth, Geo., Farming, S. 28, T. Essex, P. O. Maple Rapids, 1860.
Hinman, J., Farming, S. 16, T. Eagle, P. O. Eagle, 1839.
Hoff, James, Propr. Hotel & Barroom, Eagle, 1846.
Holbrook, J. J., Farming, S. 8, T. Duplain, P. O. Elsie, 1871.
Hollister, C. E., Farming, S. 13, T. Victor, P. O. Laingsburg, 1848.
Holmes, Geo., Farming, S. 3, T. Dewitt, P. O. Dewitt, 1856.
Howarth, J., Farming, S. 15, T. Essex, P. O. Maple Rapids, 1868.
Howe, Jeremiah B., Farming, Twp. Supervisor & School Assessor, S. 21, T. Olive, P. O. St. Johns, 1854.
Howe, Daniel, Farming, S. 11, T. Olive, P. O. St. Johns, 1861.
Hubbard, Martin D., Spring Manufacturer, St. Johns, 1868.
Hufnagel, A., Farming, S. 23, T. Dallas, P. O. Fowler, 1862.
Hugus, W. P., Farming, S. 1, T. Bingham, P. O. St. Johns, 1867.
Hulse, M. E. (Clark & Hulse), Clothiers & Gents Furnishers, St. Johns, 1885.
Hulse, J. H., Farming, S. 5, T. Duplain, P. O. Eureka, 1860.
Hunt, Frank, Farming & Stock, S. 33, T. Bath, P. O. Haslett, 1893.
Hunt, J. A., Farming & Stock, S. 13, T. Bengal, P. O. St. Johns, 1854.
Hunt, Seth, Farming, S. 28, T. Bingham, P. O. St. Johns, 1857.
Hunter, Wm. G., Farming & Stock, S. 23, T. Ovid, P. O. Ovid, 1858.
Hurd, Robert, Farming, S. 20, T. Dewitt, P. O. Lansing, 1848.
Huyck, A., Farming, S. 3, T. Victor, P. O. Victor, 1854.

Ingersoll, Eric, Farming, S. 15, T. Greenbush, P. O. St. Johns, 1857.
Irwin, Wellington, S. 24, T. Lebanon, P. O. Maple Rapids, 1847.
Ives, William, Farming & Stock, S. 31, T. Victor, P. O. Rew, 1853.
Ives, Wm., Farming, S. 36, T. Olive, P. O. Rew, 1854.

Jacob, August, Farming & Stock, S. 11, T. Riley, P. O. St. Johns, 1856.
Jaquish, A. E., Farming, S. 10, T. Bengal, P. O. St. Johns, 1854.
Jennison, H. H., Farming, S. 27, T. Eagle, P. O. Eagle, 1842.
Jennison, W. F., Farming, S. 22, T. Eagle, P. O. Eagle, 1838.
Jenkins, W. S., Farming, S. 31, T. Westphalia, P. O. Portland, 1860.
Jennie, N. J., Farming, S. 22, T. Essex, P. O. Maple Rapids, 1863.
Johnson, C. A., Insurance & Real Estate Dealer, Ovid, 1882.
Johnson, Geo. J., Editor and Propr. Elsie Enterprise, Elsie, 1895.
Jolly, P. T., Justice of the Peace, S. 11, T. Dallas, P. O. Fowler, 1854.
Jones, Leroy C., Farming & Stock, S. 6, T. Bengal, P. O. Maple Rapids, 1849.
Jones, A. E., Farming, S. 27, T. Riley, P. O. South Riley, 1848.
Jones, I. C., Farming, S. 27, T. Bengal, P. O. St. Johns, 1856.
Jones, H. F., Farming, S. 29, T. Eagle, P. O. Eagle, 1846.
Jones, H. L., Farming, S. 34, T. Eagle, P. O. Grand Ledge, 1848.

Karker, Frank, Farming, S. 13, T. Bingham, P. O. St. Johns, 1866.
Karber, Geo., Farming, S. 13, T. Bingham, P. O. St. Johns, 1866.
Kathe, Fred., Farming, S. 11, T. Bingham, P. O. St. Johns, 1867.
Kelch, G. B., Farming, S. 15, T. Bengal, P. O. St. Johns, 1855.
Kelly, Chas., Farming, S. 35, T. Westphalia, P. O. Wacousta, 1851.
Kelly, Chester, Farming, S. 35, T. Westphalia, P. O. Wacousta, 1851.
Kesby, Walter, Farming & Stock, S. 8, T. Ovid, P. O. Shepardsville, 1866.
Kesby, Henry, Farming, S. 7, T. Ovid, P. O. Shepardsville, 1875.
Kesby, Lewis, Farming, S. 8, T. Ovid, P. O. Shepardsville, 1866.
Keusch, Julius M., Farming & Stock, S. 31, T. Westphalia, P. O. Westphalia, 1854.
Keusch, Lawrence, Insurance, Westphalia, 1854.
Keys, H. N., Cashier First National Bank, Ovid, 1876.
King, B. F., Farming, S. 31, T. Watertown, P. O. Lansing, 1848.
King, W., Farming, S. 22, T. Watertown, P. O. Lansing, 1848.
Kinley, Frank, Farming, S. 35, T. Lebanon, P. O. Fowler, 1867.
Kinmond, Marguerite, Assistant Register of Deeds, St. Johns, 1877.
Kinney, B. F., Farming, S. 25, T. Watertown, P. O. North Lansing, 1847.
Knapp, M. L., Grocer, S. 11, T. Riley, P. O. Dewitt, 1862.
Knapp, W. H. H., Farming, S. 13, T. Riley, P. O. Dewitt, 1845.
Kneeland, H. B., Farming & Stock, S. 1, T. Bengal, P. O. St. Johns, 1853.
Kneeland, B. F., Retired, St. Johns, 1843.
Kneeland, Edwin, Farming, S. 13, T. Bengal, P. O. St. Johns, 1844.
Knickerbocker, Larry E., Proprietor Hotel, Maple Rapids, 1898.
Knight, David, Farming, S. 1, T. Dewitt, P. O. Rew, 1845.
Knight, E., Farming, S. 13, T. Greenbush, P. O. Eureka, 1869.
Koshi, A. H., Farming, S. 17, T. Ovid, P. O. Shepardsville, 1859.
Koster, John, Jr., Stonemason & Plasterer, Westphalia, 1870.
Koster, John, Farming, S. 28, T. Westphalia, P. O. Westphalia, 1870.
Kroll, Geo., Farming, S. 11, T. Bingham, P. O. St. Johns, 1864.
Kruger, Fred., Farming, S. 11, T. Dallas, P. O. Fowler, 1861.
Kyes, J. A., Farming, S. 35, T. Duplain, P. O. Ovid, 1879.

LaClear, L. W., Farming, S. 1, T. Greenbush, P. O. Eureka, 1860.
Laduc, H. B., Propr. Park Lake Hotel, S. 28, T. Bath, P. O. Bath, 1895.
Lake, Geo. E., Professional Horse Trainer (Lake & Phillips), also buys and sells Carriage & Road Horses, St. Johns, 1895.
Lake, A. L., Farming, S. 9, T. Bingham, P. O. St. Johns, 1849.
Lamb, C. E., Farming, S. 17, T. Bingham, P. O. St. Johns, 1847.
Lamb, W. E., Farming, S. 7, T. Bingham, P. O. St. Johns, 1862.
Lamberson, Riley, Farming & Stock, S. 17, T. Olive, P. O. St. Johns, 1866.
Lamberson, J. C., Farming, S. 17, T. Olive, P. O. St. Johns, 1866.
Langeor, Adelbert, Farming, S. 2, T. Ovid, P. O. Shepardsville, 1854.
Lapham, G. E., Farming, S. 1, T. Bingham, P. O. St. Johns, 1869.
Larkins, James, Farming, S. 28, T. Bingham, P. O. St. Johns, 1852.
Lawrence, E. D., Farming & Stock, S. 1, T. Dewitt, P. O. Rew, 1874.
Laughlin, C., Farming, S. 26, T. Bengal, P. O. St. Johns, 1846.
Lee, Elmer, Farming, S. 2, T. Duplain, P. O. Elsie, 1884.
Leeland, H. W., Farming & Stock, S. 3, T. Victor, P. O. Victor, 1886.
Lehman, Joseph, Farming, S. 27, T. Westphalia, P. O. Westphalia, 1841.
Leland, W. M., Sheriff, St. Johns, 1866.
Lenneman, John, Farming, S. 16, T. Westphalia, P. O. Westphalia, 1849.
Lerg, C. P., Farming & Stock, S. 3, T. Dewitt, P. O. Dewitt, 1856.
Letts, Walter, Farming, S. 3, T. Duplain, P. O. Elsie, 1865.
Levy, Allen, Farming, S. 15, T. Duplain, P. O. Elsie, 1866.
Lewis, Frank, Farming & Stock, S. 24, T. Duplain, P. O. Elsie, 1865.
Lewis, L., Farming & Stock, S. 24, T. Duplain, P. O. Elsie, 1865.
Lewis, I., Farming, S. 14, T. Lebanon, P. O. Maple Rapids, 1864.
Lewis, Harlem, Farming, S. 26, T. Duplain, P. O. Elsie, 1866.
Link, John, Bakery & Restaurant, Twp. Clerk, Ovid, 1898.
Lieman, A. D., Farming, S. 1, T. Duplain, P. O. Elsie, 1850.
Lieman, W. A., Farming, S. 1, T. Duplain, P. O. Elsie, 1850.
Lion, L. N., Teacher Instrumental Music, Dewitt, 1887.
Litweiler, Jacob, Farming & Stock, S. 31, T. Greenbush, P. O. St. Johns, 1863.
Livermore, H. A., Farming, Highway Commissioner, S. 22, T. Dewitt, P. O. Gunnisonville, 1867.
Lott, Albert, Farming & Stock, S. 12, T. Dewitt, P. O. Dewitt, 1838.
Lott, Benj., Farming, S. 14, T. Dewitt, P. O. Gunnisonville, 1853.
Lott, C., Farming, S. 14, T. Dewitt, P. O. Gunnisonville, 1857.
Loomis, A. J., Farming, S. 33, T. Duplain, P. O. Shepardsville, 1866.
Loomis, Ezra, Farming, S. 33, T. Duplain, P. O. Shepardsville, 1866.
Loomis, Theo., Farming, S. 16, T. Olive, P. O. St. Johns, 1856.
Lounsbury, W. N., Farming, S. 1, Ovid, P. O. Shepardsville, 1845.
Lowell, O. W., Farming, S. 23, T. Watertown, P. O. North Lansing, 1839.
Lyon, B. R., Farming, S. 8, T. Eagle, P. O. Portland, 1844.
Lyon, E. H., Farming, S. 35, T. Bengal, P. O. St. Johns, 1854.
Lyon, H. M., Farming, S. 24, T. Essex, P. O. St. Johns, 1864.

Main, John L., Farming & Stock, S. 21, T. Dewitt, P. O. North Lansing, 1896.
Makley, F. K., Farming, S. 18, T. Bath, P. O. Bath, 1854.
Mallory, C., Farming, S. 12, T. Essex, P. O. Union Home, 1845.
Mankey, Chas., Farming, S. 2, T. Dallas, P. O. Fowler, 1857.
Mankey, John, Farming, S. 10, T. Dallas, P. O. Fowler, 1856.
Mankey, John, Farming, S. 10, T. Dallas, P. O. Fowler, 1862.
Mankey, Henry, Farming, S. 33, T. Bengal, P. O. Riley, 1866.
Mankey, James, Farming, S. 24, T. Duplain, P. O. Elsie, 1867.
Manzer, C. H., Drugs & Groceries, Wacousta, 1891.
Marshall, Geo. R., County Clerk of Clinton County, St. Johns, 1868.
Marshall, Thos. J., Farming & Stock, S. 20, T. Duplain, P. O. Duplain, 1876.
Marshall, Dennis, Farming, S. 17, T. Greenbush, P. O. St. Johns, 1865.
Marshall, R., Farming, S. 30, T. Ovid, P. O. St. Johns, 1871.
Martin, Chas., Farming, S. 3, T. Lebanon, P. O. Maple Rapids, 1860.
Martin, Fred., Farming, S. 8, T. Bengal, P. O. St. Johns, 1860.
Mason, R. G., Merchant, Wacousta, 1852.
McCabe, H. D., Postmaster, St. Johns, 1865.
McCollum, I. D., Farming, S. 31, T. Ovid, P. O. St. Johns, 1843.
McCrumb, A. J., Farming, Eagle, 1845.
McCrumb, G. W., Sr., Farming, S. 16, T. Eagle, P. O. Eagle, 1846.
McGonigal, D., Farming & Stock, S. 7, T. Bath, P. O. Bath, 1851.
McIntyre, Ellsworth, Farming, S. 15, T. Olive, P. O. St. Johns, 1858.
McKee, E. C., Farming & Stock, S. 25, T. Victor, P. O. Laingsburg, 1861.
McKnight, Lewis, Farming, S. 25, T. Greenbush, P. O. Duplain, 1869.
McMaster, R., Farming, S. 5, T. Greenbush, P. O. Union Home, 1864.
Mead, Lewis C., Register of Deeds, St. Johns, 1865.
Mead, I. J., Farming, S. 16, T. Duplain, P. O. Elsie, 1855.
Merrihew, Hugh, Farming, S. 15, T. Olive, P. O. St. Johns, 1836.
Merihew, F., Farming, S. 36, T. Bingham, P. O. St. Johns, 1850.
Merrihew, M. J., Farming, S. 9, T. Olive, P. O. St. Johns, 1836.
Merrill, Chas. M., Judge of Probate, St. Johns, 1857.
Messler, Wm., Farming, S. 12, T. Essex, P. O. St. Johns, 1840.
Miller, D. F., Livery, Feed & Sale Stables, St. Johns, 1891.
Miller, Antony, Farming, S. 28, T. Westphalia, P. O. Westphalia, 1850.
Miller, John, Farming, S. 34, T. Westphalia, P. O. Westphalia, 1843.
Miller, Theo., Farming, S. 8, T. Riley, P. O. Westphalia, 1867.
Moldenhauer, C. L., Farming & Stock, S. 9, T. Watertown, P. O. Wacousta, 1866.
Monroe, Wm., Farming, S. 27, T. Bath, P. O. Bath, 1836.
Monroe, Josiah, Farming, S. 7, T. Eagle, P. O. Portland, 1836.
Monroe, Horace, Farming, S. 5, T. Eagle, P. O. Portland, 1836.
Moon, Sylvester, Farming, S. 8, T. Dewitt, P. O. Dewitt, 1844.
Moore, P., Farming & Stock, S. 32, T. Duplain, P. O. Duplain, 1857.
Moore, Richard, Farming & Sec. & Treas. The Farmers' Mutual Fire Ins. Co. of Clinton & Gratiot Counties, S. 8, T. Bingham, P. O. St. Johns, 1854.
Moore, W. H., Farming & Stock, S. 4, T. Duplain, P. O. Elsie, 1854.
Moore, J. H., Farming, S. 2, T. Duplain, P. O. Elsie, 1853.
Morgan, Wm. W., Farming & Stock, S. 30, T. Riley, P. O. Wacousta, 1866.
Morrison, D. S., Farming & Twp. Supervisor, S. 6, T. Greenbush, P. O. Union Home, 1883.
Moses, A. B., Farming & Stock, S. 1, T. Watertown, P. O. Dewitt, 1862.
Munger, O. W., President State Bank, St. Johns, 1857.
Munro, G. J., Dealer in Agricultural Implements, St. Johns, 1861.
Murdock, F. F., Marble & Granite & Cut Stone for Building, St. Johns, 1883.
Merwin, Harvey, Farming, S. 20, T. Lebanon, P. O. Matherton, 1850.
Myers, Mathew, Farming, S. 2, Ovid, P. O. Ovid, 1860.

Naerman, John, Farming, S. 22, T. Westphalia, P. O. Westphalia, 1848.
Nau, Constantin, Contractor & Builder, S. 4, T. Westphalia, P. O. Westphalia, 1866.
Neller, Valentine, Farming, S. 27, T. Olive, P. O. Rew, 1863.
Nelson, S., Farming, S. 1, T. Olive, P. O. St. Johns, 1846.
Nicholas, M., Farming, S. 35, T. Ovid, P. O. Ovid, 1863.
Nicholas, E. D., Farming, S. 35, T. Watertown, P. O. North Lansing, 1846.
Norris, D. C., Farming, S. 32, T. Bingham, P. O. St. Johns, 1841.
Norton, W. A., (Spaulding, Norton & Weimer), Prosecuting Attorney, St. John's, 1854.
Nuffer, George, Farming & Stock, S. 34, T. Lebanon, P. O. Fowler, 1855.
Nuffer, J. J., Farming & Stock, S. 3, T. Riley, P. O. St. Johns, 1851.
Nuffer, Peter, Farming & Stock, S. 33, T. Bengal, P. O. St. Johns, 1853.

Ocobock, D. H., Farming, S. 9, T. Bengal, P. O. St. Johns, 1843.
Ormston, W. S., Farming, S. 6, T. Ovid, P. O. St. Johns, 1869.
Osborn, F. M., Law & Insurance, St. Johns, 1855.
Oosterhout, E., Farming, S. 3, T. Ovid, P. O. Ovid, 1871.
Outcalt, J. W., Farming, S. 3, T. Olive, P. O. St. Johns, 1854.
Owen, J. C., Farming, S. 18, T. Riley, P. O. Riley, 1845.
Owen, J. F., Farming, S. 12, T. Lebanon, P. O. Maple Rapids, 1878.

Padgett, James, Farming, S. 5, T. Greenbush, P. O. Union Home, 1870.
Page, Jackson, Farming, S. 6, T. Duplain, P. O. Eureka, 1853.
Palmer, F., Farming, S. 32, T. Bingham, P. O. St. Johns, 1856.
Palmer, Nelson, Farming, S. 32, T. Bingham, P. O. St. Johns, 1850.
Parker, G. W., Farming, S. 5, T. Victor, P. O. Victor, 1843.
Parker, Newell, Farming, S. 10, T. Victor, P. O. Laingsburg, 1860.
Parks, Samuel, Farming, S. 17, T. Bengal, P. O. Fowler, 1851.
Parmenter, S., Supervisor & Farmer, S. 10, T. Ovid, P. O. Shepardsville, 1863.
Parr, Henry, Farming, S. 26, T. Essex, P. O. St. Johns, 1861.
Parr, S. M., Capitalist, Maple Rapids, 1865.
Partlow, John, Farming, S. 31, T. Eagle, P. O. Grand Ledge, 1846.
Partlow, Levi, Farming, S. 32, T. Eagle, P. O. Grand Ledge, 1849.
Patrick, W. C., Farming & Stock, S. 5, T. Eagle, P. O. Eagle, 1857.
Patrick, Wm., Farming, S. 5, T. Eagle, P. O. Eagle, 1852.
Patterson, H. J., Attorney, St. Johns, 1856.
Patterson & Smith, Attorneys & Counsellors at Law, Fowler, 1856.
Pearl, S. F., Manager Ovid Carriage Company, Manfrs. & Wholesale Dealers in Carriages, Buggies, Farm Wagons, etc., First-class Work Guaranteed, Ovid, 1850.
Pease, L. H., Farming, S. 17, T. Essex, P. O. Maple Rapids, 1868.
Peck, A. M., Farming, S. 34, T. Bingham, P. O. St. Johns, 1856.
Perkins, Chas., Farming, S. 22, T. Victor, P. O. Laingsburg, 1860.
Perkins, Daniel, Farming, S. 6, T. Duplain, P. O. Eureka, 1857.
Perkins, G. W., Farming, S. 5, T. Duplain, P. O. Eureka, 1855.
Perkins, J. Y., Farming, S. 22, T. Victor, P. O. Laingsburg, 1846.
Perrin, P. K. (Perrins & Baldwin), Law, Real Estate & Loans, St. Johns, 1860.
Perrin, Henry M. (Perrins & Baldwin), Law, Real Estate & Loans, St. Johns, 1857.
Perry, J. W., Farming, S. 15, T. Bengal, P. O. St. Johns, 1868.
Phelps, Grant, Farming, S. 32, T. Bingham, P. O. St. Johns, 1844.
Phillips, J. W. (Lake & Phillips), Professional Horse Trainer, Buys & Sells Carriage & Road Horses, St. Johns, 1895.
Pierce, B. L., Court Commissioner, St. Johns, 1864.
Pierce, Jas., Farming, S. 10, T. Dallas, P. O. Fowler, 1860.
Pierce, J. Q., Farming, S. 21, T. Greenbush, P. O. St. Johns, 1875.
Pike, D. S., Farming, S. 25, T. Riley, P. O. Dewitt, 1853.
Pillmore, Wm., Farming, S. 12, T. Bingham, P. O. St. Johns, 1851.
Pingel, John, Farming, S. 20, T. Riley, P. O. Riley, 1866.
Pinkey, John, Farming, S. 12, T. Lebanon, P. O. Maple Rapids, 1858.
Pinkey, Peter, Farming, S. 12, T. Lebanon, P. O. Maple Rapids, 1858.
Platte, Ferdinand, Farming, S. 17, T. Westphalia, P. O. Westphalia, 1843.
Pline, Mathias, Drover & Live Stock Dealer, Westphalia (Res. Ionia Co.).
Plowman, D. G., Farming, S. 13, T. Riley, P. O. Dewitt, 1858.
Plowman, J. E., Farming, S. 22, T. Bengal, P. O. St. Johns, 1857.
Pope, C. G., Farming, S. 17, T. Bengal, P. O. Fowler, 1866.
Post, H. A., Farming & Stock, S. 9, T. Victor, P. O. Victor, 1880.
Post, H. L., Farming, S. 8, T. Greenbush, P. O. Union Home, 1855.
Post, L. H., Manufacturing, S. 8, T. Greenbush, P. O. Union Home, 1865.
Potter, H. A., President First National Bank, Ovid.
Pratt, G. W., Farming, S. 8, Greenbush, P. O. Union Home, 1860.
Price, Chas., Farming, S. 12, T. Olive, P. O. Price, 1856.
Putnam, J. W., Farming, S. 30, T. Ovid, P. O. Shepardsville, 1873.

Rademacher, Wm. A., Farming & Stock, S. 3, T. Westphalia, P. O. Westphalia, 1871.
Rademacher, John, Farming, S. 3, T. Westphalia, P. O. Westphalia, 1866.
Reed, John, (Walker Bros. & Co.), Milling, Feed, Wholesale and Retail, Bath.
Reed, Del., (Reed and Co.), Manfrs. of Galv. Iron Washing Machines, Eagle, 1849.
Reed, M. V., Farming & Stock, S. 15, T. Riley, P. O. Oral, 1875.
Reed, James, of Reed & Co., Manufacturers, Eagle, 1868.
Reynolds, Simeon, Farming, S. 13, T. Dewitt, P. O. Dewitt, 1853.
Rice, O. R., Farming, S. 18, T. Bingham, P. O. St. Johns, 1845.
Rice, F. B., Farming, S. 19, T. Bingham, P. O. St. Johns, 1866.
Richards, W., Farming, S. 30, T. Bengal, P. O. Fowler, 1856.
Richardson, Theo., Farming, S. 12, T. Dewitt, P. O. Bath, 1831.
Richmond, H. L., Farming & Milling, S. 29, T. Greenbush, P. O. St. Johns, 1850.
Ridenour, D. W., Farming, S. 7, T. Bingham, P. O. St. Johns, 1858.
Ridenour, Daniel, Farming, S. 5, T. Bingham, P. O. St. Johns, 1852.
Robb, A., Farming, S. 31, T. Greenbush, P. O. St. Johns, 1860.
Rockwell, C. T., Farming, S. 30, T. Olive, P. O. Dewitt, 1862.
Rose, Wm. H., Treasurer Clinton County, St. Johns, 1844.
Rose, J., Farming, S. 22, T. Bath, P. O. Bath, 1852.
Rose, S. W., Farming, S. 23, T. Bath, P. O. Bath, 1838.
Rosekrans, W. A., Farming & Stock, S. 19, T. Greenbush, P. O. St. Johns, 1892.
Rossow, C., Farming, S. 7, T. Bengal, P. O. Fowler, 1854.
Rouse, Albert, Hotel Clinton, Dewitt, 1853.
Rouse, Henry, Farming, S. 17, T. Dewitt, P. O. Dewitt, 1846.
Rummell, C. M., Farming, S. 1, T. Duplain, P. O. Elsie, 1865.
Rummell, Henry, Farming, S. 10, T. Duplain, P. O. Elsie, 1865.

Sage, Wm. H., Farming & Stock, Highway Comr., S. 4, T. Dallas, P. O. Fowler, 1854.
Salisbury, J. P., Farming, S. 1, T. Dallas, P. O. Fowler, 1859.
Sargeant, M., Farming, S. 21, T. Dallas, P. O. Fowler, 1844.
Sauer, Herman, Farming & Stock, S. 4, T. Watertown, P. O. South Riley, 1893.
Schafer, J. H., Stock Buyer & Propr. Sample Room, Westphalia, 1855.
Schafer, John J., Saw Mill Owner, S. 24, T. Dallas, P. O. Fowler, 1894.
Schafer, John (Second), Farming, S. 8, T. Westphalia, P. O. Westphalia, 1846.
Schafer, N., Farming, S. 8, T. Westphalia, P. O. Westphalia, 1849.

Schneider, Andrew, Farming, S. 6, T. Westphalia, P. O. Westphalia, 1850.

Schofield, Frank, Livery, Sale & Feed Stables, St. Johns, 1868.

Schuller, John, Farming, S. 20, T. Dallas, P. O. Westphalia, 1844.

Schultz, C., Farming, S. 6, T. Bengal, P. O. Fowler, 1860.

Scott, Geo. W., Farming & Stock, S. 5, T. Dewitt, P. O. DeWitt, 1866.

Scott, Nelson, Farming & Stock, S. 15, T. Greenbush, P. O. St. Johns, 1855.

Scott, L. S., Farming, S. 2, T. Lebanon, P. O. Maple Rapids, 1858.

Scott, A., Farming, S. 21, T. Olive, P. O. St. Johns, 1852.

Scott, S. E., Farming, S. 5, T. Dewitt, P. O. Dewitt, 1838.

Scriven, John, Farming, S. 18, T. Bingham, P. O. St. Johns, 1866.

Seeley, G. S., Farming, S. 34, T. Duplain, P. O. Duplain, 1865.

Selden, J. O., Attorney at Law & Notary Public, St. Johns, 1865.

Selden, Benj., Farming, S. 18, T. Eagle, P. O. Eagle, 1851.

Servis, Wm. J., Supt. County Farm, S. 28, T. Bingham, P. O. St. Johns, 1870.

Sevy, O. B., Farmer & Mechanic, S. 23, T. Greenbush, P. O. St. Johns, 1839.

Sevy, C., Farming, S. 33, T. Greenbush, P. O. Eureka, 1869.

Severance, Lewis, Justice of Peace, St. Johns, 1865.

Shadduck, D., Farming, S. 19, T. Watertown, P. O. Grand Ledge, 1839.

Shafer, M., Farming, S. 33, T. Lebanon, P. O. Fowler, 1864.

Sharp, W. L., Farming & Stock, S. 25, T. Eagle, P. O. Grand Ledge, 1894.

Shaver, J. C., Propr. Hotel, S. 28, T. Victor, P. O. Geary.

Shaver, Theron, Farming & Stock, S. 30, T. Ovid, P. O. St. Johns, 1871.

Shaw, W. R., Sec. & Treas. Ovid Elevator Co., Grain, Seeds, Beans & Wool, Ovid, 1868.

Shepard, A. G., Attorney & Counsellor at Law, Ovid, 1873.

Shepard, B. M., Farming & Stock, Shepardsville, 1861.

Shepard, E. J., Farming & Stock, S. 18, T. Eagle, P. O. Eagle, 1864.

Shepard, Perry, Farming & Stock, S. 18, T. Eagle, P. O. Eagle, 1861.

Sherman, Cyrus, Farming & Stock, S. 22, T. Ovid, P. O. Ovid, 1865.

Shraft, Jacob, Postmaster, Fowler, 1856.

Shurburn, Geo., Farming, S. 36, T. Duplain, P. O. Ovid, 1852.

Siefert, Geo. E., Farming & Stock, S. 3, T. Bengal, P. O. St. Johns, 1859.

Silvernail, L., Farming, S. 27, T. Greenbush, P. O. St. Johns, 1861.

Simon, Anthony, Farming, S. 11, T. Westphalia, P. O. Westphalia, 1841.

Simon, August, Farming, S. 32, T. Dallas, P. O. Westphalia, 1858.

Simpson, Geo. W., Farming, S. 34, T. Ovid, P. O. Ovid, 1854.

Sirvins, Geo., Farming, S. 36, T. Bingham, P. O. St. Johns, 1855.

Slater, F. A., Farming, S. 1, T. Riley, P. O. St. Johns, 1875.

Sleight, J., Farming, S. 6, T. Bath, P. O. Bath, 1856.

Smith, A. T., Manfr. of Carriages & Wagons, St. Johns, 1863.

Smith, Clark, Farming & Stock, S. 29, T. Waterman, P. O. Grand Ledge, 1862.

Smith, E. L., (See Patterson & Smith.)

Smith, Geo. W., Farming & Stock, Twp. Supervisor, S. 33, T. Bath, P. O. Haslett, 1870.

Smith, J. Millard, Publisher, Maple Rapids, 1895.

Smith, Wm. M., Attorney & Counsellor at Law, Dewitt, 1870.

Smith, Chas., Farming, S. 12, T. Essex, P. O. St. Johns, 1862.

Smith, Cyrus, Farming, S. 33, T. Bingham, P. O. St. Johns, 1855.

Smith, E., Farming, S. 28, T. Ovid, P. O. Shepardsville, 1864.

Smith, Edward, Farming, S. 36, T. Bingham, P. O. St. Johns, 1848.

Smith, H. D., Farming, S. 17, T. Bingham, P. O. St. Johns, 1853.

Smith, I. R., Farming, S. 13, T. Dewitt, P. O. Bath, 1840.

Smith, James, Farming, S. 36, T. Bath, P. O. Haslett, 1856.

Smith, John, Farming, S. 18, T. Westphalia, P. O. Westphalia, 1846.

Smith, Peter, Farming, S. 14, T. Bingham, P. O. St. Johns, 1864.

Smith, P. H., Farming, S. 18, T. Westphalia, P. O. Westphalia, 1852.

Smith, Samuel, Farming, S. 22, T. Dewitt, P. O. Gunnisonville, 1850.

Smith, Wm., Farming, S. 8, T. Westphalia, P. O. Westphalia, 1846.

Snelling, W. H., Cashier Fowler State Savings Bank, Fowler, 1870.

Snelling, W., Farming, Elsie, 1863.

Snitgen, John, Propr. Sample Room & Dealer in Agricultural Implements, Westphalia, 1859.

Snitgen, Joseph, Farm Implements, Westphalia, 1855.

Snitgen, A., Merchant, Westphalia, 1861.

Snitgen, Frank, Clerk, Westphalia, 1865.

Soule, G. H., Farming & Stock, S. 15, T. Essex, P. O. Maple Rapids, 1860.

Soule, D. H., Farming, S. 13, T. Ovid, P. O. Ovid, 1855.

Soule, W. T., Farming, S. 14, T. Essex, P. O. Maple Rapids, 1862.

Soules, Walter, Farming, S. 31, T. Essex, P. O. Maple Rapids, 1856.

Sowls, Horace, Farming & Stock, S. 28, T. Essex, P. O. St. Johns, 1865.

Spaulding, F. M., Dealer in General Hardware, St. Johns, 1861.

Spaulding, Geo. O. L., Attorney (Spaulding, Norton & Weimer), St. Johns, 1857.

Sperry, Lyman, Farming, S. 34, T. Bingham, P. O. St. Johns, 1850.

Sperry, Wm., Farming, S. 7, T. Olive, P. O. St. Johns, 1857.

Spink, S. R., Farming, S. 9, T. Ovid, P. O. Shepardsville, 1868.

Spitzley, Anthony, Farming, S. 6, T. Westphalia, P. O. Westphalia, 1846.

Spragu- ----, Farming, S. 17, T. Victor, P. O. Laingsburg, 1852.

Springsteen, Marvin, Farming & Stock, S. 2, T. Dewitt, P. O. Dewitt, 1863.

Starkweather, Scott, Farming, S. 23, T. Duplain, P. O. Ovid, 1856.

St. Clair, P. J., Postmaster & Merchant, Jessie, 1864.

St. Clair, Howard, Farming, S. 20, T. Ovid, P. O. Shepardsville, 1839.

St. Clair, Perry, Farming, S. 20, T. Ovid, P. O. Shepardsville, 1859.

Stead, Samuel, Farming, S. 1, T. Essex, P. O. Union Home, 1860.

Stead, Wm., Farming, S. 1, T. Essex, P. O. Union Home, 1860.

Steadman, T. P., Manager Campbell, Keys & Steadman, Bankers, Elsie, 1889.

Stebbins, J. C., Farming, S. 36, T. Ovid, P. O. Ovid, 1878.

Steel, Robert G. (St. Johns Manufacturing Co.), Manfr. Tables, Etc., St. Johns, 1867.

Steel, Robert M., Capitalist & Manfr. (St. Johns Mfg. Co.), St. Johns.

St. Johns Mercantile Co., Wholesale and Retail Groceries, Dry Goods, Boots & Shoes, Thomas Bromley, Manager, St. Johns.

Stone, J. B., Farming & Teaching, S. 19, T. Riley, P. O. Westphalia, 1861.

Stowell, Clark E., Farming & Stock, S. 4, T. Bengal, P. O. Fowler, 1853.

Stowell, E., Farming & Stock, S. 25, T. Greenbush, P. O. Eureka, 1889.

Stowell, L. D., Farming & Stock, S. 10, T. Dewitt, P. O. Dewitt, 1840.

Stowell, D. K., Farming, S. 14, T. Bengal, P. O. St. Johns, 1858.

Streeter, M. T., Merchant, General Merchandise, S. 9, T. Riley, P. O. Riley, 1894.

Strickland, E. T., Farming, S. 26, T. Ovid, P. O. Ovid, 1858.

Strong, W. J., Farming & Stock, S. 1, T. Riley, P. O. Oral, 1860.

Stump, John J., Manager Hotel, Sample Room & Livery, Westphalia, 1869.

Stump, Wm., Farming, S. 5, T. Westphalia, P. O. Westphalia, 1857.

Sturgis, A. T., Farming, S. 34, T. Olive, P. O. Dewitt, 1851.

Sturgis, Daniel, Farming, S. 15, T. Bengal, P. O. St. Johns, 1864.

Sutton, H., Justice of the Peace, Fowler, 1868.

Sutfen, D. A., Drainage Commissioner, Ovid, 1853.

Sutton, W. A., Farming, S. 35, T. Bengal, P. O. Oral, 1865.

Swagart, Geo., Farming, S. 11, T. Bengal, P. O. St. Johns, 1846.

Swagart, W. A., Farming, S. 28, T. Bengal, P. O. St. Johns, 1844.

Swain, O. B., Farming, S. 21, T. Bingham, P. O. St. Johns, 1856.

Swarthout, Wm., Farming, S. 36, T. Ovid, P. O. Ovid, 1837.

Sweeney, James, Brick & Tile Manfr., Bath, 1868.

Taber, H. D., Farming, S. 33, T. Bingham, P. O. St. Johns, 1853.

Taber, W., Farming, S. 33, T. Bingham, P. O. St. Johns, 1854.

Talsor, L. C., Farming, S. 31, T. Greenbush, P. O. St. Johns, 1865.

Talsor, L. G., Farming, S. 31, T. Greenbush, P. O. St. Johns, 1842.

Taft, Stephen, Farming & Fine Stock, S. 13, T. Ovid, P. O. Jessie, 1869.

Taft, I. F., Farming, S. 25, T. Bingham, P. O. St. Johns, 1857.

Taylor, C. C., Justice of the Peace & Notary, Ovid, 1855.

Taylor, Samuel, Farming, S. 16, T. Ovid, P. O. Shepardsville, 1866.

Teed, T., Farming, S. 25, T. Bingham, P. O. St. Johns, 1863.

Teel, W. A., Merchant, General Merchandise, Bath.

Teeter & DuBois, Hillside Jersey Stock Farm, S. 29, T. Duplain, P. O. Duplain.

Terwilliger, Frank, Farming, S. 14, T. Lebanon, P. O. Maple Rapids, 1855.

Thelen, C., Farming, S. 29, T. Bengal, P. O. Fowler, 1861.

Thelen, Mat., Farming, S. 6, T. Westphalia, P. O. Westphalia, 1847.

Thiell, John, Farming, S. 36, T. Bengal, P. O. St. Johns, 1865.

Thomas, Marion, Farming & Stock, S. 23, T. Eagle, P. O. Grand Ledge, 1851.

Thomas, Geo., Farming, S. 17, T. Greenbush, P. O. St. Johns, 1850.

Thome, Peter, Loans, and Dealer in Agricultural Implements, Westphalia, 1844.

Thome, John, Farming, S. 33, T. Westphalia, P. O. Westphalia, 1860.

Thompson, Stephen, Farming, S. 30, T. Ovid, P. O. Jessie, 1864.

Throop, Calvin, Farming, S. 34, T. Watertown, P. O. Delton, 1847.

Thurston, Nathaniel, Farming, S. 15, T. Bengal, P. O. St. Johns, 1857.

Tillotson, M. L., Horticulture, S. 11, T. Duplain, P. O. Elsie, 1860.

Trancheil, J. L., Farming, S. 1, T. Bingham, P. O. St. Johns, 1865.

Travis, Chas., Farming, S. 15, T. Bengal, P. O. St. Johns, 1856.

Tripp, E. A., Farming, S. 15, T. Greenbush, P. O. St. Johns, 1862.

Trowbridge, M. F., Farming, S. 8, T. Olive, P. O. St. Johns, 1857.

Trumble, Eph., Farming, S. 1, T. Bath, P. O. Laingsburg, 1855.

Tucker, Joseph, Farming, S. 7, T. Olive, P. O. St. Johns, 1857.

Ulrich, Jas., Farming, S. 3, T. Dallas, P. O. Fowler, 1867.

Ulrich, Peter, Farming, S. 2, T. Dallas, P. O. Fowler, 1863.

Upton, H. L., Farming & Stock, S. 10, T. Victor, P. O. Victor, 1856.

Upton, Josiah, Dealer in Lumber, St. Johns, 1856.

Upton, F. R., Farming, S. 10, T. Victor, P. O. Victor, 1855.

Vance, E. E., Township Supervisor, S. 30, T. Lebanon, P. O. Matherton, 1863.

Vancousant, A. J., Propr. Sample Room, St. Johns, 1870.

Vanconsant, Peter, Propr. Sample Room, St. Johns, 1860.

Van Dyke, E., Farming, S. 13, T. Riley, P. O. St. Johns, 1869.

Van Gleson, John, Farming, S. 20, T. Bengal, P. O. St. Johns, 1846.

VanHousen, Wm., Farming, S. 28, T. Essex, P. O. St. Johns, 1860.

Van Vleet, Philo, Farming, S. 9, T. Olive, P. O. St. Johns, 1854.

Vaughan, C. C., Publisher Clinton Republican, St. Johns, 1869.

Vincent, Abraham, Farming, S. 36, T. Duplain, P. O. Ovid, 1839.

Walbridge, Henry, Attorney at Law, St. Johns, 1856.

Walbridge, H. E. (Fedewa & Walbridge), Attorney & Counsellor at Law, St. Johns, 1856.

Walker Bros. & Co., Flour Mill, Feed, Wholesale & Retail, Bath, 1890.

Walker, W. J. (See Walker Bros. & Co.), Bath.

Walker, M. D. (See Walker Bros. & Co.), Bath.

Walker, M. J., Farming, S. 34, T. Bengal, P. O. St. Johns, 1863.

Walker, W., Farming, S. 6, Bengal, P. O. Maple Rapids, 1866.

Walworth, P. E., Cashier St. Johns National Bank, St. Johns, 1865.

Waring, Clark, Farming, S. 15, T. Duplain, P. O. Elsie, 1858.

Warner, W. F., Insurance & Notary Public, Maple Rapids, 1875.

Warren, Isaac, Farming, S. 31, T. Essex, P. O. Maple Rapids, 1845.

Waters, H. H., Farming & Stock, S. 25, T. Ovid, P. O. Ovid, 1848.

Waters, H. F., Farming, S. 8, T. Ovid, P. O. Shepardsville, 1849.

Watson, J. A., Farming, S. 21, T. Duplain, P. O. Elsie, 1844.

Webb, E. M., Merchant, General Merchandise, Dewitt, 1848.

Webster, Alonzo, Farming & Stock, S. 8, T. Bath, P. O. Bath, 1867.

Webster, Mrs. C. D., Farming & Stock, S. 20, T. Riley, P. O. Riley, 1863.

Webster, Lemuel, Farming & Stock, S. 20, T. Riley, P. O. Riley, 1863.

Webster, Andrew, Farming, S. 3, T. Bingham, P. O. St. Johns, 1837.

Webster, E. B., Farming, S. 27, T. Essex, P. O. St. Johns, 1869.

Weimer, W. F., (Spaulding, Norton & Weimer), Atty. & Counsellor at Law, St. Johns, 1880.

Wellinsen, William, Farming & Stock, S. 22, T. Bingham, P. O. St. Johns, 1867.

Weller, Andrew, Farming & Stock, S. 2, Bengal, P. O. St. Johns, 1847.

West, A. S., Farming & Stock, Twp. Supervisor, S. 15, T. Dewitt, P. O. Gunnisonville, 1867.

West, Samuel, Farming, S. 5, T. Greenbush, P. O. Union Home, 1865.

Whitaker, Albert, Farming & Stock, Twp. Treasurer, S. 36, T. Riley, P. O. Dewitt, 1855.

Whitaker, E., Farming, S. 23, T. Bengal, P. O. St. Johns, 1866.

Whitaker, O. C., Farming, S. 36, T. Bengal, P. O. St. Johns, 1856.

White, Henry, Farming, S. 13, T. Lebanon, P. O. Maple Rapids, 1859.

White, Isaac, Farming, S. 13, T. Lebanon, P. O. Maple Rapids, 1859.

Whittacre, J. M., Farming, S. 1, T. Lebanon, P. O. Maple Rapids, 1860.

Wieber, Anthony, Farming & Stock, S. 13, T. Westphalia, P. O. Westphalia, 1843.

Wieber, J. J., Farming, S. 13, T. Westphalia, P. O. Westphalia, 1871.

Wieber, E. P., Farming, S. 13, T. Westphalia, P. O. Westphalia, 1874.

Wieber, Joseph, Painter & Paper Hanger, Westphalia, 1861.

Wilcox, Earl, Farming, S. 36, T. Greenbush, P. O. St. Johns, 1840.

Wilcox, L. C., Farming, S. 25, T. Bingham, P. O. St. Johns, 1857.

Wilkins, Geo. R., (Wilkins Bros.), Livery, Sale & Feed Stables, St. Johns, 1847.

Wilkins, Joseph, (Wilkins Bros.), Livery, Sale & Feed Stables, St. Johns, 1847.

Wilkinson, Levi, Farming, S. 15, T. Eagle, P. O. Eagle, 1840.

Williams, C. S., Farming & Stock, S. 1, T. Bingham, P. O. St. Johns, 1865.

Williams, S. H., Merchant & Blacksmith, S. 8, T. Riley, P. O. Riley, 1884.

Williams, F. J., Farming, S. 16, T. Ovid, P. O. Shepardsville, 1854.

Williams, C. A., Farming, S. 1, T. Ovid, P. O. Ovid, 1857.

Williams, G. B., Farming, S. 34, T. Bingham, P. O. St. Johns, 1860.

Williams, N., Farming, S. 35, T. Bingham, P. O. St. Johns, 1855.

Williams, W. A., Farming, S. 1, T. Bingham, P. O. St. Johns, 1865.

Williams, C., Farming, S. 7, T. Bingham, P. O. Fowler, 1854.

Winger, A., Farming, S. 36, T. Bengal, P. O. St. Johns, 1860.

Winfield, J. J., Farming, S. 3, T. Ovid, P. O. Shepardsville, 1857.

Winston, R. M., School Commissioner, St. Johns, 1894.

Wolcott, J. A., Farming & Stock, S. 35, T. Essex, P. O. St. Johns, 1853.

Wood, Thomas, Farming, S. 12, T. Eagle, P. O. Wacousta, 1856.

Woodaus, W., Farming, S. 9, T. Victor, P. O. Victor, 1846.

Woodman, J. C., Propr. Hotel, Bath.

Woodruff, Geo., Merchant, Boots & Shoes, St. Johns, 1869.

Wooll, I. A., Propr. Oak Grove Fruit Farm, Specialty, Strawberries & Raspberries, S. 2, T. Duplain, P. O. Elsie, 1869.

Worden, James, Farming, S. 4, T. Victor, P. O. Victor, 1854.

Yariger, Samuel, Farming, S. 25, T. Ovid, P. O. Jessie, 1865.

Young, B. F., Farming & Stock, S. 3, T. Bengal, P. O. St. Johns, 1844.

Young, Isaac, Farming & Stock, S. 20, T. Essex, P. O. St. Johns, 1854.

Young, J., Farming, S. 24, T. Essex, P. O. St. Johns, 1860.

Yonry, W. F., Farming & Teaching, S. 22, T. Bath, P. O. Bath, 1869.

ANALYSIS OF THE SYSTEM

OF

UNITED STATES LAND SURVEYS

METES AND BOUNDS.

UP to the time of the Revolutionary War, or until about the beginning of the present century, land, when parcelled out, and sold or granted, was described by "Metes and Bounds," and that system is still in existence in the following States, or in those portions of them which had been sold or granted when the present plan of surveys was adopted, viz.: New York, Pennsylvania, New Jersey, Delaware, Maryland, Virginia, North and South Carolina, Georgia, Tennessee, Kentucky, Texas, and the six New England States. To describe land by "Metes and Bounds," is to have a known land-mark for a place of beginning, and then follow a line according to the compass-needle (or magnetic bearing), or the course of a stream, or track of an ancient highway. This plan has resulted in endless confusion and litigation, as land-marks decay and change, and it is a well-known fact that the compass-needle varies and does not always point due North.

As an example of this plan of dividing lands, the following description of a farm laid out by "Metes and Bounds," is given: "Beginning at a stone on the Bank of Doe River, at a point where the highway from A. to B. crosses said river (see point marked C. on Diagram 1); thence 40° North of West 100 rods to a large stump; thence 10° North of West 90 rods; thence 15° West of North 80 rods to an oak tree (see Witness Tree on Diagram 1); thence due East 150 rods to the highway; thence following the course of the highway 50 rods due North; thence 5° North of East 90 rods; thence 45° East of South 60 rods; thence 10° North of East 300 rods to the Doe River; thence following the course of the river Southwesterly to the place of beginning." This, which is a very simple and moderate description by "Metes and Bounds," would leave the boundaries of the farm as shown in Diagram 1.

MERIDIANS AND BASE LINES.

DIAGRAM 2.

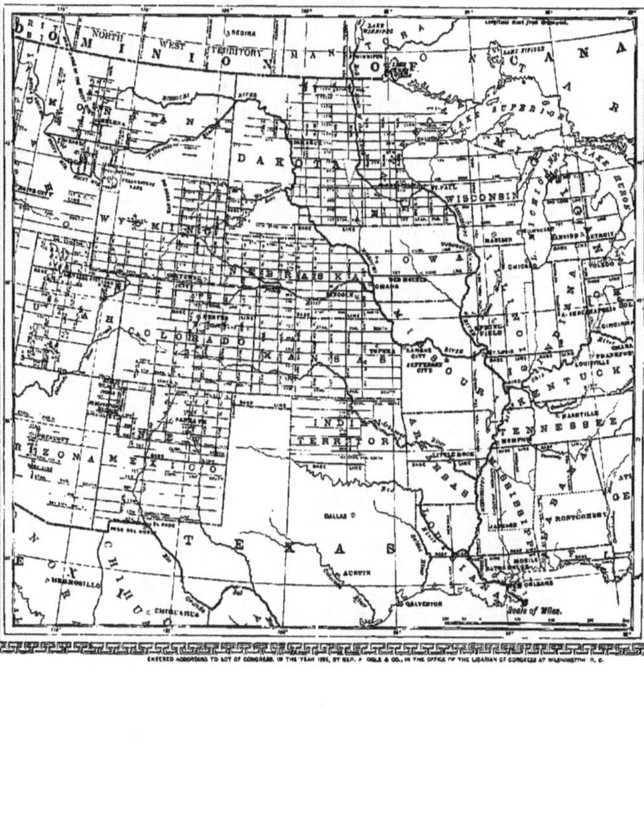

THE present system of Governmental Land Surveys was adopted by Congress on the 7th of May, 1785. It has been in use ever since and is the legal method of describing and dividing lands. It is called the "Rectangular System," that is, all its distances and bearings are measured from two lines which are at right angles to each other, viz.: +. These two lines, from which the measurements are made, are the Principal Meridians, which run North and South, and the Base Lines, which run East and West. These Principal Meridians are established, with great accuracy, by astronomical observations. Each Principal Meridian has its Base Line, and these two lines form the basis or foundation for the surveys or measurement of all the lands within the territory which they control.

Diagram 2 shows all of the Principal Meridians and Base Lines in the central portion of the United States, and from it the territory governed by each Meridian and Base Line may be readily distinguished. Each Meridian and Base Line is marked with its proper number or name, as are also the Standard Parallels and guide (or auxiliary) Meridians.

Diagram 3 illustrates what is meant when this method is termed the "Rectangular System," and how the measurements are based on lines which run at right angles to each other. The heavy line running North and South (marked A. A.) represents the Principal Meridian, in this case say the 5th Principal Meridian. The heavy line running East and West (marked B. B.) is the Base Line. These lines are used as the starting points or basis of all measurements or surveys made in territory controlled by the 5th Principal Meridian. The same fact applies to all other Principal Meridians and their Base Lines. Commencing at the Principal Meridian, at intervals of six miles, lines are run North and South, parallel to the Meridian. This plan is followed both East and West of the Meridian throughout the territory controlled by the Meridian.

UNITED STATES LAND SURVEYS.

These lines are termed "Range Lines." They divide the land into strips or divisions six miles wide, extending North and South, parallel with the Meridian. Each division is called a Range. Ranges are numbered from one upward, commencing at the Meridian; and their numbers are indicated by Roman characters. For instance, the first division (or first six miles) west of the Meridian is Range I. West; the next is Range II. West; then comes Range III., IV., V., VI., VII., and so on, until the territory governed by another Principal Meridian is reached. In the same manner the Ranges East of the Meridian are numbered, the words East or West being always used to indicate the direction from the Principal Meridian. See Diagram 3.

Commencing at the Base Line, at intervals of six miles, lines are run East and West parallel with the Base Line. These are designated as Township Lines. They divide the land into strips or divisions six miles wide, extending East and West, parallel with the Base Line. This plan is followed both North and South of the Base Line until the territory governed by another Principal Meridian and Base Line is reached. These divisions or Townships are numbered from one upward, both North and South of the Base Line, and their numbers are indicated by figures. For instance: The first six mile division north of the Base Line is Township 1 North; the next is Township 2 North; then comes Township 3, 4, 5, and 6, North, and so on. The same plan is followed South of the Base Line; the Townships being designated as Township 1 South, Township 2 South, and so on. The "North" or "South" (the initials N. or S. being generally used) indicates the direction from the Base Line. See Diagram 3.

These Township and Range Lines, crossing each other, as shown in Diagram 3, form squares, which are called "Townships" or "Government Townships," which are six miles square, or as nearly that as it is possible to make them. These Townships are a very important feature in locating or describing a piece of land. The location of a Government Township, however, is very readily found when the number of the Township and Range is given, by merely counting the number indicated from the Base Line and Principal Meridian. As an example of this, Township 8 North, Range 4, West of the 5th Principal Meridian, is at once located on the square marked ★ on Diagram 3, by counting eight tiers north of the Base Line and 4 tiers west of the Meridian.

DIAGRAM 3

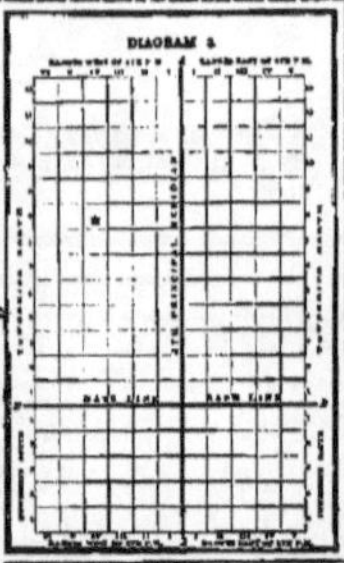

TOWNSHIPS OF LAND.

TOWNSHIPS are the largest subdivisions of land run out by the United States Surveyors. In the Governmental Surveys Township Lines are the first to be run, and a Township Corner is established every six miles and marked. This is called "Townshipping." After the Township Corners have been carefully located, the Section and Quarter Section Corners are established. Each Township is six miles square and contains 23,040 acres, or 36 square miles, as near as it is possible to make them. This, however, is frequently made impossible by: (1st) the presence of lakes and large streams; (2nd) by State boundaries not falling exactly on Township Lines; (3rd) by the convergence of Meridians or curvature of the earth's surface; and (4th) by inaccurate surveys.

Each Township, unless it is one of the exceptional cases referred to, is divided into 36 squares, which are called Sections. These Sections are intended to be one mile, or 320 rods, square and contain 640 acres of land. Sections are numbered consecutively from 1 to 36, as shown on Diagram 4. Beginning with Section 1 in the Northeast Corner, they run West to 6, then East to 12, then West to 18, and so on, back and forth, until they end with Section 36 in the Southeast Corner.

Diagram 4 shows a plat of a Township as it is divided and platted by the government surveyors. These Townships are called Government Townships or Congressional Townships, to distinguish them from Civil Townships or organized Townships, as frequently the lines of organized Townships do not conform to the Government Township lines.

DIAGRAM 4

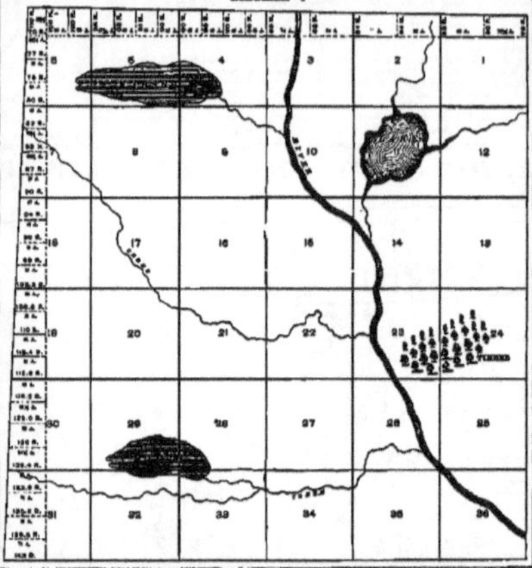

SECTIONS OF LAND.

DIAGRAM 5 illustrates how a section may be subdivided, although the Diagram only gives a few of the many subdivisions into which a section may be divided. All Sections (except fractional Sections) are supposed to be 320 rods, or one mile, square and therefore contain 640 acres—a number easily divisible. Sections are subdivided into fractional parts to suit the convenience of the owners of the land. A half-section contains 320 acres; a quarter-section contains 160 acres; half of a quarter contains 80 acres, and quarter of a quarter contains 40 acres, and so on. Each piece of land is described according to the portion of the section which it embraces—as the Northeast quarter of Section 10; or the Southeast quarter of the Southeast quarter of Section 10. Diagram 5 shows how many of these subdivisions are platted, and also shows the plan of designating and describing them by initial letters as each parcel of land on the Diagram is marked with its description.

As has already been stated, all Sections (except Fractional Sections which are explained elsewhere) are supposed to contain 640 acres, and even though mistakes have been made in surveying, as is frequently the case, making sections larger or smaller than 640 acres, the Government recognizes no variation, but sells or grants each regular section as containing 640 acres "more or less."

The Government Surveyors are not required to subdivide sections by running lines within them, but they usually establish Quarter Posts on Section Lines on each side of a section at the points marked A. B. C. and D. on Diagram 5. After establishing Township corners, Section Lines are the next to be run, and section corners are established. When these are carefully located the Quarter Posts are located at points as nearly equidistant between Section Corners as possible. These corners when established by Government Surveyors cannot be changed, even though it is conclusively shown that mistakes have been made which cause some sections or quarter sections to be either larger or smaller than others. The laws, however, of all the States provide certain rules for local surveyors to follow in dividing Sections into smaller parcels of land than has been outlined in the Governmental surveys. For instance, in dividing a quarter section into two parcels, the distance between the Government Corners is carefully measured and the new post is located at a point equidistant between them. This plan is followed in running out "eighties," "forties," "twenties," etc. In this way, if the Government division overruns or falls short, each portion gains or loses its proportion. This is not the case, however, with Fractional Sections along the North or West sides of a Township, or adjoining a lake or large stream.

DIAGRAM 5.

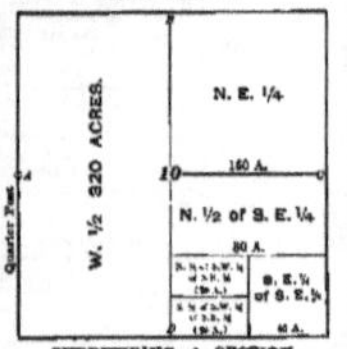

SUBDIVIDING A SECTION.

FRACTIONAL PIECES OF LAND.

CONGRESSIONAL Townships vary considerably as to size and boundaries. Mistakes made in surveying and the fact that Meridians converge as they run North cause every Township to vary more or less from the 23,040 acres which a perfect Township would contain. See Diagram 4. In arranging a Township into Sections all the surplus or deficiency of land is given to, or taken from, the North and West tiers of Sections. In other words, all Sections in the Township are made full—640 acres—except those on the North and West, which are given all the land that is left after forming the other 25 Sections.

Diagram 4 illustrates how the surplus or deficiency is distributed and the Sections it affects. It will be seen that Sections 1, 2, 3, 4, 5, 6, 7, 18, 19, 30 and 31, are the "Fractional Sections," or the Sections which are affected if the Township overruns or falls short. Inside of these Fractional Sections, all of the surplus or deficiency of land (over or under 640 acres) is carried to the "forties" or "eighties" that touch the Township Line. These pieces of land are called "Fractional Forties" or "Fractional Eighties," as the case may be. Diagrams 4 and 6 show the manner of marking the acreage and outlining the boundaries of these "Fractions."

Diagram 6 illustrates how the surplus or deficiency of land inside of these Sections is distributed and which "forties" or "eighties" it affects. From this arrangement it will be seen that in any Section that touches the North or West Township Lines, the Southeast Quarter may be full—160 acres—while another quarter of the same Section may be much larger or smaller. Frequently these fractional "forties" or "eighties" are lotted as shown in Diagram 6. They are always described as fractional tracts of land, as the "fractional S. W. ¼ of Section 6," etc. Of course those portions of these Sections which are not affected by these variations are described in the usual manner—as Southeast ¼ of Section 6. As a rule Townships are narrower at the North than at the South side. The Meridians of Longitude (which run North and South) converge as they run North and South from the Equator. They begin at the Equator with a definite width between them and gradually converge until they all meet at the poles. Now, as the Range lines are run North and South, it will at once be seen that the convergence of Meridians will cause every Congressional Township (North of the Equator) to be narrower at its North than at its South side, as stated. See Diagram 4. In addition to this fact, mistakes of measurement are constantly and almost unavoidably made in running both Township and Range lines, and if no new starting points were established the lines would become confused and unreliable, and the size and shape of Townships materially affected by the time the surveys had extended even a hundred miles from the Base Line and Principal Meridian. In order to correct the surveys and variations caused by the difference of latitude and straighten the lines, "Correction Lines" (or Guide Meridians and Standard Parallels) are established at frequent intervals, usually as follows: North of the Base Line a Correction Line is run East and West parallel with the Base Line, usually every twenty-four miles. South of the Base Line a Correction Line is usually established every thirty miles. Both East and West of the Principal Meridian "Correction Lines" are usually established every 48 miles. All Correction Lines are located by careful measurement, and the succeeding surveys are based upon them.

DIAGRAM 6.

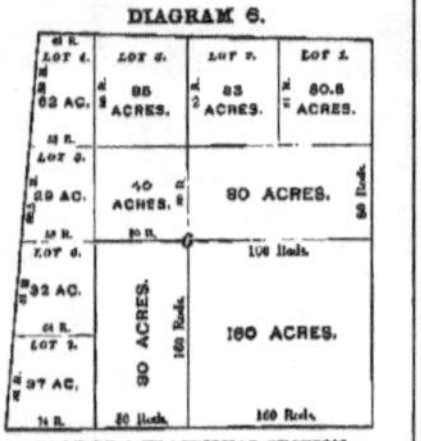

PLAT OF A FRACTIONAL SECTION.

DIGEST OF THE SYSTEM OF CIVIL GOVERNMENT,

WITH A REVIEW OF THE DUTIES AND POWERS OF THE PRINCIPAL OFFICIALS CONNECTED WITH THE VARIOUS BRANCHES OF NATIONAL, STATE, COUNTY AND TOWNSHIP GOVERNMENT.

NATIONAL GOVERNMENT.

THE GOVERNMENT of the United States is one of limited and specific powers, strictly outlined and defined by a written constitution. The constitution was adopted in 1787, and, with the amendments that have since been made, it forms the basis of the entire fabric of government under which we live. The constitution created three distinct branches of government, each of which is entirely separate and distinct from the others. They are the executive, legislative and judicial departments. The constitution specifically vests the executive power in the President, but all members of the cabinet are usually classed with the executive department; the legislative power is held by Congress, and the judicial authority is vested in the Supreme Court and various other courts which Congress has provided for in pursuance of the provisions of the constitution.

It has been the aim of these pages to explain each of these different branches of government, and to briefly review the duties and powers of the principal officials connected with each department.

The President and Vice-President are elected by popular vote, but the vote of each State is separate, so that a candidate may have a large majority of the aggregate popular vote of the country and yet fail to be elected. The Presidential election is held on the first Tuesday after the first Monday in November, when Presidential electors are chosen in and for the various States, each State having as many electors as it has representatives in both branches of Congress. The electors are chosen by the ballots of the people of their States, and all the electors of a State constitute an electoral college. The electors meet in each State at the capital on the first Wednesday in December following a National election and vote for President and Vice-President, certificates of which are forwarded to the President of the Senate, at Washington, who, on the second Wednesday in February opens the certificates and counts the votes in the presence of both Houses of Congress and declares the result; and the final step is the inauguration, which takes place on the 4th of March. The law provides that if neither of the candidates have a majority then the House of Representatives shall elect a President from the three candidates receiving the highest electoral vote. In elections of this kind each State is entitled to only one vote, and two-thirds of the States form a quorum.

PRESIDENT OF THE UNITED STATES.

The President is the highest executive officer of the United States. He is elected for the term of four years, and receives a salary of $50,000 per annum. He must be thirty-five years old or more, and a native-born citizen of the United States. The President is charged with a general supervision over the faithful execution of laws passed by Congress, and has supervision over all executive departments of the government. He appoints a Cabinet of eight officials who become the heads of the various departments, and these departments are intended to be managed and conducted as the President directs. The President is Commander-in-Chief of the Army and Navy. He has power to grant pardons and reprieves for all offenses against the United States, except in cases of impeachment; has power, with the advice and consent of the Senate, to make treaties. He nominates, and with the advice and consent of the Senate, appoints Ambassadors and other public Ministers and Consuls, all Judges of the United States courts, and all other executive officers of the United States, except in such cases where the appointments may be vested in the various "departments." When the Senate is not in session he can appoint, subject to its action when it reassembles. He has power, in certain extraordinary occasions, to call together both Houses of Congress, or either of them, in extra session; and is required from time to time to communicate with Congress, as to the state of the Union, and offer such suggestions or recommendations as he may deem proper. He is empowered to approve or veto all measures adopted by Congress, but it is provided that any measure may be passed over his veto by a two-thirds vote of Congress.

The President consults frequently with his Cabinet, and nearly all important official matters are discussed by that body. In case the office of President becomes vacant through the death, removal or resignation of the incumbent, the law provides that the office shall in turn be filled by the Vice-President, Secretary of State, and other Cabinet Ministers in regular order.

VICE-PRESIDENT.

The Vice-President of the United States is elected for the term of four years, and receives a salary of $10,000. In case of the death, removal or resignation of the President, the Vice-President succeeds him. The chief duty of the Vice-President is to act as the presiding officer of the Senate. He has no vote in the Senate, except in cases of a tie, or an equal division of the members of that body. The Vice-President administers the oath of office to the Senators.

STATE DEPARTMENT.

The head of this department is the Secretary of State, who is appointed by the President as a member of the Cabinet, and receives a salary of $8,000 per year. The law provides that in case the office of President becomes vacant, through the death, removal or resignation of both the President and Vice-President, the Secretary of State assumes the duties of the Presidency. The Secretary of State may be said to be the official Secretary of the President, and countersigns all commissions issued by the President.

The Secretary of State is the head of the Department of State and is the chief diplomatic officer of the United States. In his department and under his supervision is conducted the public business relating to foreign affairs; to correspondence, commissions or instructions to or with public Ministers from the United States; or to negotiations with Ministers from foreign States; or to memorials or other applications from foreigners, or foreign public Ministers, or citizens of this country in foreign lands, or complications arising therefrom. The Secretary of State also has charge of all other business connected with foreign affairs, extradition matters and diplomatic officers; furnishing passports to vessels going to foreign countries, etc., and has charge of the Great Seal of the United States.

Connected with the Department of State and forming a part of it in the great work of performing and caring for the duties outlined are the following bureaus:

The Diplomatic Bureau, which looks after the affairs pertaining to foreign governments.

The Consular Bureau, correspondence with consulates.

The Bureau of Indexes and Archives, the duties of which are to open the official mails, prepare an abstract of the daily correspondence and an index of it, and superintend miscellaneous work of department.

The Bureau of Accounts, in which all of the finances of the department are looked after, such as the custody and disbursement of appropriations; also indemnity funds and bonds; also care of the building and property of the department, etc.

The Bureau of Rolls and Library, which is charged with the custody of treaties, rolls, public documents, etc.; has care of revolutionary archives, of international commissions, superintendence of library, etc.

The Bureau of Statistics, for the preparation of reports on commercial relations.

The chiefs of all of these bureaus receive $2,100 per year. In addition to these there are connected with the State Department the offices of translator, at $2,100 per year; assistant secretary, $4,500; second assistant secretary, $3,500; third assistant secretary, $3,500; solicitor, $3,500; chief clerk, $2,750; clerk to Secretary of State, $2,000; passport clerk, $1,400. Besides these there are the various comptrollers, auditors, clerks and assistants, which number well up into the thousands.

TREASURY DEPARTMENT.

This department was organized in 1789. The head of this department, known as the Secretary of the Treasury, is appointed by the President, is a member of the Cabinet, and receives a salary of $8,000 per annum. The Treasury Department is one of the most important branches of the national government, as it has charge of the financial affairs of the government, custody of public funds, collection of revenue and maintenance of public credit. Among the many important duties devolving upon this department are the following: It attends to the collection of all internal revenues and duties on imports, and the prevention of frauds in these departments. All claims and demands, either by the United States or against them, and all the accounts in which the United States are interested, either as debtors or creditors, must be settled and adjusted in the Treasury Department. This department also includes the Bureau of the Mint, in which the government coin and moneys are manufactured. The Treasury Department authorizes the organization of national banks and has supervision over them; has charge of the coast surveys, the lighthouses, marine hospitals, etc. It has charge of all moneys belonging to the United States; designates depositories of public moneys, keeps a complete and accurate system of accounting, showing the receipts and disbursements of the Treasury, and makes reports at stated intervals showing the condition of public finances, public expenditures and the public debt.

There are a great many very important officials connected with the Treasury Department, chief among which are the following, viz.: Private secretary of the head of the department, at $2,400 per year; three assistant secretaries, at $4,500 each; chief clerk, $3,000; chief of appointment division, $2,750; chief of warrants division, $2,750; chief of public moneys division, $2,500; chief of customs division, $2,700; acting chief of revenue marine division, $2,500; chief of stationery division, $2,500; chief of loans and currency division, $2,500; chief of miscellaneous division, $2,500; supervising special agent, $8 per day; government actuary, $1,800; supervising architect, $4,500; steamboat inspector, $3,500; chief Bureau of Statistics, $3,000; life saving service superintendent, $4,000; assistant, $2,500; commissioner Bureau of Navigation, $3,000; superintendent United States coast and geodetic survey, $6,000; supervising surgeon-general marine hospital service, $4,000; Bureau of Engraving and Printing, chief, $4,500; assistant chief, $2,250; superintendent engraving division, $3,600.

The foregoing will serve to show many of the lines of work attended to in the Treasury Department, as the names of these offices explain the branch of work they are charged with attending to. There are a number of other important offices in the department that should be mentioned, among them being the following:

The Solicitor of the Treasury, or chief attorney, who receives $4,500 per year for attending to the legal matters connected with the department.

The Commissioner of Customs, who receives $4,000 per year and his deputy $2,250, has charge of all accounts of the revenue from customs and disbursements, and for the building and repairing of custom houses.

The Treasurer of the United States receives $6,000 per year, assistant treasurer $3,600, and superintendent of national banks (Red. Div.) $3,500. The Treasurer receives and keeps the government funds, either at headquarters or in the Sub-Treasuries or government depositories, paying it out upon warrants drawn in accordance with the law, and pays all interest on the national debt.

The Register of the Treasury is paid a salary of $4,000 per year, and his assistant $2,250. The Register keeps the accounts of public expenditures and receipts; receives the returns and makes out the official statements of United States commerce and navigation; receives from first comptroller and Commissioner of Customs all accounts and vouchers acted on by them and files the same.

The Comptroller of the Currency receives $5,000 per year and his deputy $2,800. This bureau is charged with a general supervision of the national banks and matters connected with the issuing of paper money.

The Director of the Mint receives $4,500 per annum, and is charged with a general supervision over all the coinage of the government.

Comptrollers. The first and second comptrollers are paid a salary of $5,000 per year, and each of their deputies receive $2,700. The first comptroller revises and certifies the accounts of the civil and diplomatic service and public lands. The second comptroller revises and certifies the accounts of the army and navy and of the Pension and Indian Bureaus.

Auditors. There are six auditors connected with the Treasury Department, each of whom receives a salary of $3,600 per year, and is allowed a deputy at a salary of $2,250 per annum. No one auditor takes rank over another. The first auditor receives and adjusts the accounts of the revenue and disbursements, appropriations and expenditures on account of the civil list and under special acts of Congress, reporting the balances to the commissioners of the customs and first comptroller respectively for their decision. The second auditor devotes most of his attention to army affairs; looks after all the accounts relating to the pay, clothing and recruiting of the army; the arsenals, armories and ordnance; all accounts relating to the Indian Department; reporting to the second comptroller. The third auditor has all accounts for subsistence of the army, military academy, military roads, fortifications, quartermaster's department, certain pensions, claims arising for military service previous to 1817; for all property lost in the military service; he reports also to the second comptroller. The fourth auditor also reports to the second comptroller, and attends to all accounts of the service connected with the navy. The fifth auditor reports to the first comptroller, and adjusts all accounts connected with the diplomatic service of the Department of State. The sixth auditor adjusts all accounts growing from the service of the Post Office Department.

WAR DEPARTMENT.

The War Department was organized in August, 1789. The head of this department is known as the Secretary of War; is appointed by the President, and receives a salary of $8,000 per annum. The War Department attends to the execution of all laws affecting the Regular Army, and carries out and performs such duties as may be provided for by law or directed by the President relative to military forces, military commissions and the warlike stores of the United States. In former years this department also had charge of Indian as well as military affairs, but this has been transferred to the Department of the Interior. The War Department is also required, among other duties, to maintain the signal service and provide for taking meteorological observations at various points on the continent, and give telegraphic notice of the approach of storms. There is also maintained a Civil Engineering Department, through the aid of which is carried out such improvements in rivers and harbors as may be authorized by Congress. The Secretary of War also has supervision over the West Point Military Academy.

The private clerk for the head of the War Department is paid $2,000 per year; assistant secretary, $4,500; chief clerk, $2,750. The most of the subordinates and assistants in the War Department, except those mentioned, are officers of the Regular Army, who are paid salaries and perquisites.

The Commanding General comes next to the Secretary, and receives a salary of $7,500 per year. He looks after the arrangement of military forces, superintends the recruiting service and discipline of the army, orders courts-martial, and in a general sense is charged with seeing to the enforcement of the laws and regulations of the army. The Adjutant-General keeps the rolls and the orders issued. The Quartermaster-General has charge of the barracks and the supplies, etc., that may be required for the army. The Commissary-General is head of the Subsistence Department, and has supervision over the purchasing and issuing army rations. The Judge Advocate General is the head of the department of military justice. The Surgeon General, as the name implies, looks after the affairs of the army relating to sick, wounded, hospital, etc. The Paymaster-General is the disbursing officer for the money required by the department. There is also the Ordnance office, controlling ordnance stores, arsenals, armories, the manufacture of arms, etc. The Topographical office has charge of all plats and drawings of all surveys made for military purposes. Besides these there are the Inspector-General's Department and departments devoted to war records, publications, etc.

In this connection it may be of interest to the general reader to refer briefly to a few facts concerning the Regular Army. The United States is divided for this purpose into a number of military districts. The head of each department receives his general instructions and orders from headquarters. The term of service in the Regular Army is five years. The pay of private soldiers at the start is $13 per month and rations, and this is increased according to time of service, being $21 per month and rations after twenty years' service. The pay of the officers is proportioned to their rank. Colonels receive $4,500 per year; brigadier generals, $5,500; and major generals, $7,500.

NAVY DEPARTMENT.

The head of this department is the Secretary of the Navy, who is appointed by the President, and receives a salary of $8,000 per annum. This department is charged with the duty of attending to the construction, armament, equipment and employment of vessels of war, as well as all other matters connected with naval affairs, and appropriations made therefor by Congress. The Secretary of the Navy has direct control of the United States Naval Academy at Annapolis, Maryland; issues orders to the commanders of the various squadrons; has general authority over the Marine Corps; and has control of all the several bureaus of the Navy Department.

There are a number of bureaus organized in the Navy Department for the purpose of more thoroughly handling the work, among the most important of which may be mentioned the following: Bureau of Steam Engineering; Bureau of Medicine and Surgery; Bureau of Navigation; Bureau of Provisions and Clothing; Bureau of Yards and Docks; Bureau of Ordnance; Bureau of Equipment and Recruiting; Bureau of Construction and Repair. Attached to this department are also officials or bureaus to attend to the following matters: Marine Barracks, Washington, D. C.; Museum of Hygiene; Naval Dispensary; Board of Inspection and Survey; Navy Supplies and Accounts; Naval Observatory; Hydrographic Office; Library and War Records; Naval Intelligence; Nautical Almanac, etc.

Rear-admirals in the Navy are paid $6,000 per year; commodores, $5,000; captains, $4,500; lieutenant-commanders, $3,000; medical directors (rank of captains), $4,400; medical inspectors (rank of commanders), $4,400; pay directors (rank of captains), $4,400; pay inspectors (rank of commanders), $4,400. In the Engineer Corps the chief engineers are also paid $4,400 per year.

POST OFFICE DEPARTMENT.

This is one of the most important branches of the National Government. Its head is the Postmaster-General, who is appointed by the President, and receives a salary of $8,000 per annum. The Post Office Department has supervision over the execution of all laws passed by Congress affecting the postal service, and has general supervision over everything relating to the gathering, carrying and distribution of United States mails; superintends the distribution and disposal of all moneys belonging to, or appropriated for, the department; and the instruction of and supervision over all persons in the postal service, with reference to their duties.

In providing for handling the general work of the Post Office Department it has been found necessary to create four bureaus, or offices, as they are termed, each of which is presided over by an assistant postmaster-general, who each receive $4,000 per annum; are all subject to the direction and supervision of the head of the department. A review of these various bureaus and their principal officials, with the name of the office, will show very clearly the work handled by each.

The first assistant postmaster-general is allowed a chief clerk at $2,000 per year; superintendent of post office supplies, $2,000; superintendent free delivery division, $3,000; chief division of salaries and allowances, $2,200; superintendent money order system, $3,500; superintendent Dead Letter Office, $2,500; chief division of correspondence, $1,800.

The second assistant postmaster-general has charge of a number of divisions, indicated by the following officials who are under his control; superintendent of railway adjustments, at $2,000 per year; chief of inspection division, $2,000; chief of mail equipment division, $1,800; general superintendent railway mail service, $3,500; superintendent foreign mails, $3,000.

The third assistant postmaster-general has charge of the postage stamp division and the finance division. The chief of the former receives $2,250 per annum, and of the latter $2,000 per year.

The fourth assistant postmaster-general has control of a number of divisions, as indicated by the following officials who are under his supervision, viz.: Chief of the division of appointments, who is paid $2,000 per annum; chief of the division of bonds and commissions, $2,000; chief post office inspector, $3,000; and the division of mail depredations.

Besides the various chiefs of divisions mentioned above there are connected with the Post Office Department a law clerk, at $2,500 per year; appointment clerk, at $1,800; assistant attorney-general, $4,000; superintendent and disbursing clerk, $2,100; and a topographer, at $2,500 per annum.

DEPARTMENT OF THE INTERIOR.

The Interior Department is under the immediate control of the Secretary of the Interior. He is appointed by the President, and receives a salary of $8,000 per year. In this department, as the name implies, is conducted most of the public business relating to domestic or internal affairs, and, like most of the other executive departments, it is divided into a number of subdivisions and branches. The Secretary of the Interior is charged with a general supervision over public business connected with the following branches, viz.: 1st. The census of the United States. 2d. All matters connected with public lands. 3d. Everything relating to the Indians or Indian affairs. 4th. All matters concerning pensions or bounty lands. 5th. The issuance and filing of patents and caveats. 6th. The custody and distribution of publications. 7th. The compilation of statistics relating to educational matters in the various States.

DIGEST OF THE SYSTEM OF CIVIL GOVERNMENT.

He also has oversight over several of the Government's charitable and benevolent institutions. For the purpose of handling properly the business connected with most of the subjects mentioned, there are bureaus organized for the purpose.

The salaries paid to the principal officials connected with the Interior Department are as follows: First assistant secretary of the interior, $4,500 per year; assistant secretary, $4,000; chief clerk, $2,750; assistant attorney-general (Dept. of Interior), $5,000; commissioner of the General Land Office, $5,000; commissioner of Indian affairs, $4,000; superintendent of Indian schools, $3,000; commissioner of the Pension Office, $5,000; medical referee, $3,000; commissioner of railroads, $4,500; commissioner of the Patent Office, $5,000; commissioner of the Education Office, $5,000; director of geological surveys, $6,000; superintendent of the Census Office, $6,000.

DEPARTMENT OF AGRICULTURE.

This department was formerly connected with the Interior Department, but in 1889 it was reorganized and made independent, and the Secretary of Agriculture was made a member of the Cabinet. The head of this department is appointed by the President, and receives a salary of $8,000 per annum.

The general duty and design of the Department of Agriculture is to acquire and diffuse among the people of the United States useful information on subjects connected with agriculture in the most general and comprehensive sense of that word, and to procure, propagate and distribute among the people new and valuable seeds and plants.

The following is a list of the chief officials connected with the Department of Agriculture and their salaries, and the list will also serve to indicate the various lines of work handled by and the various duties which devolve upon the department, viz.: Assistant secretary of agriculture receives $4,500 per annum; chief of Weather Bureau, $4,500; chief of Bureau of Animal Industry, $3,000; statistician, $2,500; chemist, $2,500; entomologist, $2,500; botanist, $2,500; ornithologist, $2,500; chief of forestry division, $2,000; pomologist, $2,500; chief of vegetable pathology division, $2,000; microscopist, $2,500; director of office of experimental stations, $25,000; chief division of accounts, $2,500; chief of division of records and editing, $2,500; chief of division of illustrations and engravings, $2,000; horticulturist, $2,500.

DEPARTMENT OF JUSTICE.

The head of the Department of Justice is the Attorney-General, who is appointed by the President, and receives a salary of $8,000 per annum. The principal assistant of the Attorney-General is the Solicitor-General, who receives $7,000 per year. There are a number of assistant attorney-generals who receive $5,000 per annum, and a special assistant attorney-general is appointed for nearly all of the various departments, including the Treasury, State, Post Office and Interior Departments. Besides these there are a number of special officials connected with the Department of Justice, such as examiner of titles, who receives $2,750 per annum; superintendent of buildings, $2,500; appointment and disbursing clerk, $2,000, and attorney in charge of pardons, $2,400.

The Attorney-General is the legal adviser of the President, and it is the duty of the Department of Justice to give all opinions and render all services requiring the skill of persons learned in the law necessary to enable the President and other officers of the various Government departments to discharge their respective duties. This department is also required to prosecute or defend all suits or proceedings in which the United States is interested. The Attorney-General has general supervision over all the solicitors for the various departments; and also exercises general superintendence and direction over all United States marshals and United States district attorneys of all the districts of the United States and Territories.

INDEPENDENT DEPARTMENTS.

There are several independent departments, which, although none of them are as important as the foregoing, and their heads are not Cabinet members, yet they form a very necessary part and attend to very important branches of the National Government.

Government Printing Office. The head of this branch of public work is the Public Printer, who is appointed by the President, and receives a salary of $4,500 per year. His chief clerk is paid $2,400 per year, and there is a foreman of printing and a foreman of binding, each of whom receive $2,100 per annum.

Civil Service Commission. This commission consists of three commissioners, each of whom are paid $3,500 per year. The chief examiner connected with the commission is paid $3,000 per annum, and the secretary $2,000.

Interstate Commerce Commission. This commission was created for the purpose, and charged with the duty, of seeing that the laws regulating interstate commerce were faithfully executed and observed, and to prevent unjust discrimination on the part of railway corporations and common carriers. The commission consists of five commissioners appointed from different sections of the United States, each of whom receives a salary of $7,500 per year. The secretary of the commission receives a salary of $3,500 per annum.

Department of Labor. The general design of this department is to collect, assort and systematize statistical details relating to the different branches of labor in the United States. The head of this department is known as the Commissioner of the Department of Labor, and he is paid a salary of $5,000 per annum. His chief clerk receives $2,500 per year, and disbursing clerk $1,800.

JUDICIARY.

The judicial powers of the United States are vested in the following-named courts, viz.: The United States Supreme Court, consisting of one chief justice and eight associate justices; the United States Court of Claims, which consists of one chief justice and four judges; the United States Circuit Court of Appeals; and the United States Circuit and District Courts. All judges of United States Courts are appointed for life, or during "good behavior." The chief justice of the United States Supreme Court receives a salary of $10,500 per annum, and the associate justices $10,000 each. The circuit judges receive a salary of $6,000 each per annum, district judges $5,000, and judges of the Court of Claims $4,500 each per year.

The jurisdiction of the United States Courts extends to all cases in law and in equity arising under the Constitution, the laws of the United States, and treaties; to all cases affecting ambassadors, other public ministers and consuls; to all cases of admiralty and maritime jurisdiction; to controversies to which the United States shall be a party; to controversies between two or more States; between a State and a citizen of another State; between citizens of different States; between citizens of the same State claiming lands under grants of different States. In all cases affecting ambassadors, other public ministers and consuls, and those in which a State is a party the Supreme Court has original jurisdiction. In the other cases the Supreme Court has appellate jurisdiction.

LEGISLATIVE DEPARTMENT.

The legislative powers of the United States are vested in a Congress, which consists of a Senate and House of Representatives, and which meets annually at Washington on the first Monday of December. The constitution gives to Congress the following general powers: To lay and collect taxes, duties, imposts and excises; pay the debts of the United States; borrow money on the credit of the United States; to regulate commerce; to establish uniform laws on naturalization and bankruptcy; to coin money and regulate the value thereof; fix the standard of weights and measures; to declare war; to raise and support armies (but it is provided that no appropriation for this purpose can be for a longer period than for two years); to provide and maintain a navy; to grant letters of marque and reprisal, and make rules concerning captures on land and water; to make rules for the government and regulation of the land and naval forces; to establish postoffices and post-roads; to promote the progress of science and the useful arts by securing for limited times, to authors and inventors, the exclusive right to their respective writings and discoveries; to constitute tribunals inferior to the Supreme Court; to define and punish piracies and felonies committed on the high seas and offenses against the law of nations; to exercise exclusive legislation over the District of Columbia and places purchased for forts, magazines, arsenals, etc.; and further to make all laws necessary for the general welfare of the United States, and for "carrying into execution the foregoing powers, and all other powers vested by the Constitution in the Government of the United States, or in any department or officer thereof." The Constitution expressly forbids Congress making any law respecting the establishment of religion, or prohibiting the free exercise thereof, or abridging the freedom of speech, or of the press, or the right of the people peaceably to assemble, and to petition the Government for a redress of grievances. Congress cannot suspend the privilege of the writ of *habeas corpus* except in cases of rebellion or invasion when the public safety may require it. No bill of attainder or *ex post facto* law can be passed. No tax or duty can be laid on articles exported from any State. No preference can be given by any regulation of commerce or revenue to the ports of one State over those of another. No title of nobility can be granted. Every law passed by Congress must be submitted to the President for his approval. If he returns it with his objections, or vetoes it, the measure may be passed over his veto by a two-thirds vote of both branches of Congress.

The Senate, or the "Upper House of Congress," is composed of two Senators from each State in the Union. They are elected by the Legislatures of their respective States, for the term of six years, and receive a salary of $5,000 per annum. No person can be elected to the United States Senate who has not attained the age of thirty years, been nine years a citizen of the United States, and is when elected an inhabitant of the State from which he is chosen. The Senate has sole power to try all impeachments. Its consent and confirmation is necessary for all important officers appointed by the President. Its consent is also necessary to conclude any treaty.

The House of Representatives is the "Lower House of Congress." Each State in the Union is divided into congressional districts, of as nearly equal population as is practicable. In each district a representative is elected by the people for a term of two years, and each is paid a salary of $5,000 per year. Besides these, a delegate from each organized Territory is admitted to the House of Representatives, who is not entitled to vote, but has the right to debate on all subjects in which the Territory which he represents has an interest. No person can be a representative who has not attained the age of twenty-five years, been for seven years a citizen of the United States, and is at the time of his election an inhabitant of the State from which he is chosen. All bills for raising revenue must originate in the House of Representatives.

STATE GOVERNMENT.

THE method of State government throughout the United States follows very closely the general plan of government that prevails in national affairs. The various functions of government in State affairs are handled in departments, with a State officer at the head of each branch, and the lines are clearly drawn between the executive, legislative and judicial powers. All the States are governed under a constitution, which outlines and defines the powers which each of these departments shall exercise and possess. All of the most important State officials are elected by the people, but in many of the States the less important offices are filled by appointment of the Governor, by and with the consent of the State Senate.

GOVERNOR.

The Governor is the highest executive officer in all the States of the Union, and is elected by a direct vote of the people. The term of office varies materially in the different States, ranging from two to six years. As to the matter of salary that the Governor receives, it also differs widely throughout the different States and is subject to frequent change. At the present writing two States—New York and Pennsylvania—pay their Governors $10,000 per year; Illinois and California both pay $6,000 per annum; Minnesota, Indiana, Kentucky, Massachusetts, Missouri, Nevada, New Jersey, Virginia and Wisconsin all pay $5,000 per year; Maryland pays $4,500; Michigan, Louisiana, Mississippi, Ohio, Tennessee and Texas pay $4,000, Florida and Arkansas pay $3,500; Alabama, Colorado, Iowa, Kansas and North Carolina all pay $3,000; West Virginia, $2,700; Montana and Washington, $2,800; the Dakotas and Nebraska, $2,500; Connecticut, Delaware and Maine, $2,000; Oregon, $1,500, and New Hampshire, Rhode Island and Vermont $1,000. About the only statement concerning the qualifications required for this office that would be common to all the States is that he must be a citizen of the State in which he is elected. In most of the States, in addition to the salary named, the Governor is furnished with a residence, which is known as the "Executive Mansion."

The powers and duties that devolve upon the Governor are about the same in all of the States. He is charged with a general supervision over the faithful execution of the laws, and is the legal custodian of all the property of the State not specifically entrusted to other officers by law, and is authorized to take summary possession of such property. He is expected to communicate by message to each session of the State legislature such information or recommendations regarding State affairs as he may deem necessary and proper, and he is empowered to call extra sessions of that body whenever the public welfare may demand. He accounts to the same body for all moneys received and paid out, and presents estimates of amounts to be raised by taxation for various purposes. He has a negative (or veto) upon all laws passed by the Legislature, but it is provided that measures may be passed over his veto by a two-thirds vote of that body. The Governor is commander-in-chief of the State military or naval forces, and has authority to call out such forces to preserve peace and execute the laws when the local authorities are unable to accomplish this. He may require the opinion of the various State officers upon any subject relating to their respective offices, and examines and approves the bonds of State officials. In many States the Governor has power to grant reprieves and pardons, after conviction, for all offenses against the State except in cases of impeachment; but in a few of the States the pardoning power is vested in a board selected for that purpose, of which the Governor is generally ex-officio a member. The Governor has the appointment of a number of State officers, and in many cases if an elective office becomes vacant he has power to fill it by appointment; has power in many States to suspend a State officer, or even a county officer, pending a legal investigation. The Governor issues requisitions upon the executives of other States for parties charged with crime who escape to other States, and he has power to issue warrants for fleeing criminals upon requisition of other Governors.

LIEUTENANT-GOVERNOR.

The office of Lieutenant-Governor does not exist in all of the States in the Union, at least not under this name, as in a few of the States this officer is only known as the President of the State Senate. In some of the States the Lieutenant-Governor is paid a certain amount per day during sessions of the Legislature or General Assembly, and in others he is allowed a fixed salary, but it is provided that if the duties of Governor should devolve upon him, he shall during the continuance of such emergency be entitled to the emoluments thereof. The principal duty of the Lieutenant-Governor is to act as the presiding officer of the State Senate or Upper House of the State Legislature. In case a vacancy should occur in the office of Governor, the Lieutenant-Governor would act as Governor until such vacancy was filled by election; and in all cases where the Lieutenant-Governor is unable to act as presiding officer of the Senate, a President *pro tempore* is chosen by that body. The Lieutenant-Governor has no vote in the Senate except in cases of a tie or equal division of the members.

SECRETARY OF STATE.

The office of Secretary of State is one of the most important offices within the gift of the people of a State, and the office exists under this name in every State in the Union. The Secretary of State may be said to be the official secretary of the Governor, and countersigns all commissions issued by the chief executive, and he is the custodian of the Great Seal of the State. As a rule it is the duty of the Secretary of State to call the House of Representatives to order and preside until a temporary presiding officer, or Speaker, is elected. It is his duty to see that halls are prepared for the Legislature or General Assembly; he prepares the legislative manual and causes it to be printed and distributed; secures the printing and distribution of the State laws; indexes and files executive documents; provides and distributes election blanks; has charge of all books, bills, papers, etc., of the Legislature, and is practically "keeper of all public acts, laws, records, bonds, etc." The Secretary of State is required to keep a register of all the official acts of the Governor, and affixes the Seal of the State to all official commissions, etc., keeps a record of them, and is obliged to give any person a copy of the same when demanded. In all of the States the Secretary of State is *ex-officio* member of a number of the official State boards, but no list of these could be given that would apply to all States, as they are different in the various States.

STATE AUDITOR.

The office of Auditor of State exists under one name or another in nearly every State in the Union. The title of this office, however, is not alike in all the States, as in many of them, notably California, Connecticut, Florida, Georgia, Maryland, Nevada, New Jersey, New York, South Carolina, Tennessee, Texas, and a few others, it is known as State Comptroller. In a few of the States, including Michigan and Pennsylvania, the office is called Auditor-General, and in two of the States the public accounts are audited by a Board of Auditors. In all the States, however, the duties that devolve upon this branch of the State government are practically the same, and a general explanation of the scope of work handled by the State Auditor in one State will apply, except as regards minor details, to all of the States. It is the duty of the State Auditor to keep the accounts of the State with any other State or Territory, and with the United States and all public officers, corporations and individuals having accounts with his State. He audits the accounts of all public officers who are to be paid out of the State Treasury, and all persons who are authorized to receive money out of the State Treasury. In fact, all claims against the State which are to be paid out of the State Treasury must be presented to the Auditor, who, after the same is adjusted, issues warrants therefor payable at the Treasury. A complete record of each warrant is kept by the Auditor, who also keeps an account with the State Treasurer, charging him with all moneys paid into the Treasury, and giving credit for all warrants paid, and the books and vouchers of the Treasury must balance therewith, as settlements are made between these two officers at stated intervals. In a number of the States the Auditor is charged with a general supervision over certain corporations, such as insurance and banking corporations and building and loan associations, and in some States is *ex-officio* a member of a number of State boards. He generally has authority to make and execute satisfactions of judgments and assignments thereof in behalf of the State.

STATE TREASURER.

This is one of the most important executive offices in the gift of the people of a State. The State Treasurer handles vast sums of the people's money, and as a rule a very heavy bond, ranging from $500,000 up into the millions, is required of him; and generally the Governor is empowered to demand additional bonds if he deems the bond insufficient to fully protect the State.

The duties of the State Treasurer are implied by the title of the office, and they are very much the same throughout all of the States of the Union. The State Treasurer is custodian of all the State funds. He deposits these funds in banks, which give bonds to secure the Treasurer or State against loss, and which pay interest on daily balances. The Treasurer pays out State funds only on warrants issued or signed by the State Auditor, or other proper official, and a full record of all warrants is kept in both the auditing office and Treasurer's office. The plan by which the Treasurer receives the revenues of the State is different in different States. In some States the Auditor issues an order for him to receive the same and charges the amount against the Treasury. In others he is charged with all moneys which he is entitled to receive, and then given credit for delinquencies. In still other States the Treasurer issues duplicate receipts for all moneys paid in, which must be countersigned by the Auditor to be valid, and one of these must be deposited with the Auditor, so he may charge the amount against the Treasurer. In this way a double system is carried on—both Auditor and Treasurer keeping a full account of all moneys received and paid out, and their books and accounts must balance, as at stated intervals the Treasurer must make settlements with the Auditor and submit books, vouchers, etc., to the Legislature. In most of the States the State Treasurer is required to publish at stated times, in the newspapers at the capital, an itemized statement of the public accounts, expenditures, funds, receipts and disbursements. He is also required to make a complete report and itemized statement to each session of the Legislature. In nearly all of the States the law is very explicit in outlining the duties of the State Treasurer, the following being very common provisions in relation to the office, viz.: That a complete record of all moneys must be kept, showing what is received or paid out of the various "funds," which "funds" must be exhibited in separate accounts. In several of the States the Governor and one or two other State officials constitute a board, which must at certain times examine and check up the accounts, books and vouchers of the State Treasurer and ascertain the amount of funds in the Treasury.

ATTORNEY-GENERAL.

The Attorney-General, as the name implies, is the general legal counsel or lawyer for the various branches of the State government. In all of the States the powers and duties of the Attorney-General are very similar. It is his duty to appear for the State in all actions and proceedings in the Supreme Court in which the State has an interest; to institute and prosecute in all courts all actions, either for or against a State officer, in which the State has an interest; to consult with and advise the various county or state's attorneys in matters relating to their official duties, and when public interest requires he assists them in criminal prosecutions. It is his duty to consult with and advise the Governor and other State officers, and give, when requested, written opinions on legal or constitutional questions relating to their official duties, and to give written opinions when requested by the Legislature or any committee thereof. It is also his duty to prepare, when necessary, drafts for contracts or other writings relating to subjects in which the State is interested. He is required to enforce the proper application of funds appropriated to the various State institutions, and prosecute breaches of trust in the administration of the same; and when necessary prosecute corporations for failure or refusal to comply with the laws; to prosecute official bonds of delinquent officers or corporations in which the State has an interest. The Attorney-General is required to keep a record of all actions, complaints, opinions, etc.

STATE SUPERINTENDENT OR SUPERINTENDENT OF PUBLIC INSTRUCTION.

This is an office which exists in nearly every State in the Union. In three or four of the States the management of the educational interests of the State is vested in a State Board of Education, but in these cases the secretary of the board assumes most of the detail work that in most of the States devolve upon the State Superintendent. The full title given to this office is not the same in all of the States, but it is generally called "State Superintendent of Public Instruction or Public Schools." In Ohio, Maine and Rhode Island, and a few others, the office is termed "Commissioner of Schools."

The duties of the State Superintendent are very much alike in all of the States, as he is charged with a general supervision over the educational interests of the State and of the public schools. In many States his authority is not limited to the public schools, and he is authorized by law to demand full reports from all colleges, academies or private schools. It is his duty to secure at regular intervals reports from all public educational institutions and file all papers, reports and documents transmitted to him by local or county school officers. He is the general adviser and assistant of the various county superintendents or school officers, to whom he must give, when requested, his written opinion upon questions arising under the school law. It is also his duty to hear and determine controversies arising under the school laws coming to him by appeal from a county superintendent or school official. He prepares and distributes school registers, school blanks, etc., and is generally given the power to make such rules and regulations as are necessary to carry into efficient and uniform effect the provisions of the laws relating to schools. The State Superintendent is required to make a detailed report to each regular session of the State Legislature, showing an abstract of the common school reports; a statement of the condition of public schools and State educational institutions; the amount of money collected and expended, and all other matters relating to the schools or school funds that have been reported to him. He is forbidden from becoming interested in the sale of any school furniture, book or apparatus.

STATE LIBRARIAN.

In nearly all of the States the laws provide for a State officer under the title of "State Librarian." As a rule the office is filled by appointment of the Governor, although in a few States it is an elective office and is filled by direct vote of the people. The State Librarian is the custodian of all the books and property belonging to the State Library, and is required to give a bond for the proper discharge of his duties and safekeeping of the property intrusted to his care, as in many of the States the State Library is an immensely important and valuable collection. In some of the States the Supreme Court judges prescribe all library rules and regulations. In others they have a Library Board of Trustees, which is sometimes made up of the Governor and certain other State officials, who constitute a board of commissioners for the management of the State Library.

ADJUTANT-GENERAL.

In nearly all of the States provision is made for an Adjutant-General, who is either elected by the people or appointed by the Governor. The name of the office implies the branch of work which is handled by its incumbent. It is the duty of the Adjutant-General to issue and transmit all orders of the Commander-in-Chief with reference to the militia or military organizations of the State. He keeps a record of all military officers commissioned by the Governor, and of all general and special orders and regulations issued, and of all other matters relating to the men, property, ordnance, stores, camp and garrison equipage pertaining to the State militia or military forces.

PUBLIC EXAMINER OR BANK EXAMINER.

This is a State office that is found in only about one-half of the States. In some States it is known as Bank Comptroller and in others the duties which devolve upon this officer are handled by a "department" in the State Auditor's office. The general duties and plan of conducting this work, in many respects, is very similar, but there is a great difference between the various States in the officers who attend to it. Where this is made a separate State office, generally speaking, the requirements are that he must be a skilled accountant and expert bookkeeper, and cannot be an officer of any of the public institutions, nor interested in any of the financial corporations which it may be his duty to examine. He is charged with the duty of visiting and inspecting the financial accounts and standing of certain corporations and institutions organized under the State laws. In several of the States it is also made his duty to visit certain county officials at stated intervals, and inspect their books and accounts, and enforce a uniform system of bookkeeping by State and county officers.

COMMISSIONER OR SUPERINTENDENT OF INSURANCE.

In all of the States of the Union the department relating to insurance has grown to be an important branch of State government. The method of controlling the insurance business differs materially in many of the States, although they are all gradually moving in the same direction, viz., creating a department or State office in which all matters relating to insurance and insurance companies are attended to. In former years, in nearly all of the States, the insurance business formed a department in the State Auditor's office, and was handled by him or his appointees. Now, however, in nearly all the Northern States and many of the Southern States, they have a separate and distinct insurance department, the head of which is either elected by the people or appointed by the Governor. The duties and powers of the insurance department of the various States are very similar. A general provision is that the head of this department must be experienced in insurance matters, and he is prohibited from holding an interest in any insurance company. The Commissioner or Superintendent of Insurance has extensive powers concerning insurance matters, and it is his duty to see that all laws respecting and regulating insurance and insurance companies are faithfully observed; he issues licenses to insurance companies, and it is his duty to revoke the license of any company not conforming to the law. Reports are made to him at stated times by the various companies, and he has power to examine fully into their condition, assets, etc. He files in his office the various documents relating to insurance companies, together with their statements, etc., and at regular intervals makes full reports to the Governor or Legislature.

COMMISSIONER OF LABOR STATISTICS.

In several of the States a "Commissioner of Labor Statistics" is appointed by the Governor, who is the head of what may be termed the labor bureau. In a great majority of the States, however, this branch of work is taken care of by a board of labor commissioners, a bureau of statistics or by the State Auditor and his appointees. The general design of this bureau or commission is to collect, assort and systematize, and present in regular reports to the Legislature, statistical details relating to the different departments of labor in the State, and make such recommendations as may be deemed proper and necessary concerning the commercial, industrial, social, educational and sanitary conditions of the laboring classes.

OTHER STATE OFFICERS.

In all of the States there exist one or more other State officers in addition to those already mentioned, which are made necessary by local condition or local business interests. It is, therefore, unnecessary to mention any of these at length in this article. It may be stated, however, that in all of the States may be found two or more of the following State officers, and further, that each one of the following-named officers is found in some State in the Union, viz.: Superintendent or commissioner of agriculture, commissioner of mines, secretary of agricultural board, secretary of internal affairs, clerk and reporter of the Supreme Court, commissioner of railways, commissioner of immigration, State printer, State binder, land agent or commissioner, commissioner, register or superintendent of State land office, register of lands, commissioner of schools and lands, surveyor-general, inspector-general, State oil inspector, dairy commissioner.

STATE BOARDS.

Besides the officers and departments which have already been mentioned, there are a number of State boards or bureaus that are necessary in carrying on the complex business connected with the government of a State. The following list of such State boards and bureaus includes all that can be found in the majority of the States; some of them, however, are only found in a few of the States, because they are of a local nature and are only made necessary by the existence of certain local conditions or business interests. It will also be observed that some of the boards named cover the same line of work that has already been mentioned as belonging to some State officer. This grows from the fact that a few of the States place the management of certain lines of work in the hands of a State board, while in others, instead of having a State board they delegate the powers and duties to a single State official. All of the States, however, have a number of the State boards mentioned in this list, the names of which imply the line of work each attends to, viz.: Railroad and warehouse commissioners, board of equalization, board or commission of agriculture, university trustees, board or commissioners of public charities, canal commissioners, penitentiary commissioners, board of health, dental examiners, trustees of historical library, board of pharmacy, commission of claims, live stock commissioners, fish commissioners, inspectors of coal mines, labor commissioners, board of education, board of public works, board of pardons, assessment commissioners.

LEGISLATURE OR GENERAL ASSEMBLY.

The law-making power of every State is termed the "Legislative Department." The legislative power, according to the constitutions of the various States, is vested in a body termed the Legislature or General Assembly, which consists of an Upper and Lower House, designated usually as the Senate and House of Representatives. In a few of the States the Lower House is called "The Assembly." In most of the States the Legislature meets in regular sessions every two years, but this is not the universal rule, as in a few of the States the law provides for annual sessions. In all of the States, however, a provision is made whereby the Governor may, on extraordinary occasions, call a special session by issuing a proclamation.

The Legislative Department has the power to pass all such laws as may be necessary for the welfare of the State, and carry into effect the provisions of the constitution. The Legislature receives the reports of the Governor, together with the reports of the various other State officers; they provide by appropriation for the ordinary and contingent expenses of the government; at regular times provided by law they apportion the State into political districts, and make all other provisions for carrying on the State government. There is a general prohibition against the passage of any *ex post facto* law, or law impairing the obligation of contracts, or making any irrevocable grant of special privileges or immunities. Any measure to become a law must be passed by both branches of the Legislature, and then be presented to the Governor for his approval. If he withholds his approval (or vetoes it), the measure may be repassed by a two-thirds vote of the Legislature, when it will become a law notwithstanding the Governor's veto.

SENATE.

The Senate is the Upper House of the Legislature or General Assembly. The various States are divided into senatorial districts, in each of which a Senator is elected—the term of office varying from two to four years. Except in three or four of the States the presiding officer of the Senate is the Lieutenant-Governor, although a President *pro tem.* is usually elected, who acts as presiding officer during the absence of the Lieutenant-Governor. The presiding officer has no vote, however, in the Senate, except when that body is equally divided. Every Senator has one vote upon all questions, and the right to be heard in advocating or opposing the passage of any measure brought before the Legislature. In filling all of the most important State offices that are to be appointed by the Governor, the appointments must be approved or confirmed by the Senate.

HOUSE OF REPRESENTATIVES.

The Lower House of the State Legislature, in nearly if not quite all the States of the Union, is termed the House of Representatives. Like the Senators, every member of the House has the right to be heard in advocating or opposing any measure brought before the body of which he is a member. The House is given the sole power of impeachment, but all impeachments must be tried by the Senate. As a general rule, there is a provision that all bills for raising revenue must originate in the House.

JUDICIARY.

The "Judicial Department" is justly regarded as one of the most important and powerful branches of government of either the State or Nation, as it becomes the duty of this department to pass upon and interpret, and thereby either annul or give validity to all the most important measures and acts of both the legislative and executive branches of the government.

It is impossible in a general article to give a detailed review or description of the construction and make-up of the judicial departments of the various States. The courts are so differently arranged both as to their make-up and jurisdiction that it would be useless to try to give the reader a general description that would accurately cover the ground.

In all of the States, except, possibly, one or two, the highest judicial authority of the State is known as the Supreme Court, and unless questions are involved which give the United States Courts jurisdiction, it is the court of last resort. The Supreme Court is made up of a chief justice and the several associate justices or judges as may be provided for by the laws of the various States, usually from four to six. Generally these officers are elected by the people, either from the State at large or (in three of the States) as representing certain districts, but this is not the case always, as in several States they are chosen by the Governor or Legislature. In all of the States the Supreme Court has appellate jurisdiction both in law and in equity, and has original jurisdiction in remedial cases, *mandamus*, *habeas corpus* and cases relating to the revenue, but there is no trial by jury in this court.

Various other courts are provided for by the laws of the different States, such as appellate courts, circuit or district courts, probate courts, county courts, superior courts, municipal courts, courts of justices of the peace, etc. The jurisdiction of all these courts is, of course, inferior to that of the Supreme Court, and varies greatly in the different States. Besides these, where there are large cities, various other courts are also established to aid in caring for the enormous amount of judicial work that arises from such vast and complex business interests. The various courts are also provided with the necessary officials for carrying on the judicial business—such as clerks of court, court reporters, bailiffs, etc.

COUNTY GOVERNMENT.

So far as the principal county offices are concerned, the general arrangement and method of handling the public business is very much the same in all of the States; but the offices are called by different names, and in minor details—such as transferring from one office to another certain minor lines of work—there are a number of points in which the method of county government in the various States differs. The writer has adopted the names of the principal county offices which are most common in the Northern States, as in the Southern and New England States there are scarcely any two States in which the names or titles of all the county offices are identical.

AUDITING OFFICE AND CLERK OF THE COUNTY BOARD.

Generally the principal auditing officer of the county is known as the "county auditor" or "county clerk." In Illinois, Kansas, Missouri, Wisconsin and many other States the office is called "county clerk." In Indiana, Iowa, Minnesota, North Dakota, Ohio and others it is termed "county auditor." In a few of the States under certain conditions this office is merged with some other county office. A notable example of this is in the State of Michigan, where they have one official, under the simple title of "clerk," who looks after about all of the work which in most of the States devolves upon both the county clerk and also clerk of court. In all of the States a bond in a moderate sum is required of the county clerk or auditor, and he is paid a salary of from $1,200 to $3,000 per year, besides in some States being allowed certain fees, unless it is in a very large and heavily populated county, where the salary paid is of necessity much higher than this amount. No county treasurer or member of the county board is eligible to this office. In general terms it may be stated as a rule the auditor acts as the clerk or secretary of the official county board, although in a few of the States the court clerk is required to look after this matter. The clerk of the county board keeps an accurate record of the board's proceedings and carefully preserves all documents, records, books, maps and papers which may be brought before the board, or which the law provides shall be deposited in his office. In the auditing office an accurate account is kept with the county treasurer. Generally they file the duplicates of the receipts given by the county treasurer, charging him with all money paid into the treasury and giving credit for all warrants paid. The *general* plan of paying claims against a county is as follows: If the claim is one in which the amount due is fixed by law, or is authorized to be fixed by some other person or tribunal, the auditor issues a warrant or order which will be paid by the treasurer, the certificate upon which it is allowed being duly filed. In all other cases the claim must be allowed by the county board, and the chairman or presiding officer issues a warrant or order which is attested by the clerk. A complete record of all these county warrants or orders is kept, and the accounts of the county treasurer must balance therewith. The above in general terms outline the most important branch of work which the county clerk or county auditor looks after in most of the States, but in all of the States the law requires him to look after a number of other matters, although in these there is no uniformity between the various States, and no general description of these minor or additional duties could be given that would apply to all the States.

COUNTY TREASURER.

This is an office which exists in all of the States, and it is one of the most important of the various offices necessary in carrying on the business of a county. It is an elective office in all of the States, and the term of office is usually either two or four years, but a very common provision in the various States is that after serving for one term as county treasurer a party shall be ineligible to the office until the intervention of at least one term after the expiration of the term for which he was elected. This provision, however, does not exist in all of the States, as in some of them the county treasurer is eligible for re-election for any number of terms.

The general duties of the county treasurers throughout the various States is very similar. The county treasurer is the principal custodian of the funds belonging to the county. It is his duty to receive and safely keep the revenues and other public moneys of the county, and all funds authorized to be paid to him, and disburse the same pursuant to law. He is required to keep proper books of account, in which he must keep a regular, just and true account of all moneys, revenues and funds received by him, stating particularly the time, when, of whom and on what fund or account each particular sum was received; and also of all moneys, revenues and funds paid out by him according to law, stating particularly the time when, to whom and on what fund payment is made from. The books of the county treasurer must always be subject to the inspection of the county board, which, at stated intervals, examines his books and makes settlements with him. In some of the States the provisions of the law relating to county treasurers are very strict; some of them provide for a county board of auditors, who are expected, several times a year, to examine the funds, accounts and vouchers of the treasury without previous notice to the treasurer; and in some it is provided that this board, or the county board, shall designate a bank (or banks) in which the treasurer is required to keep the county funds deposited—the banks being required to pay interest on daily or monthly balances and give bond to indemnify the county against loss. As a general rule the county treasurer is only authorized to pay out county funds on warrants or orders issued by the chairman of the county board and attested by the clerk, or in certain cases on warrants or orders of the county auditing office. A complete record of these warrants or orders is kept, and the treasurer's accounts must balance therewith. In most of the States the law is very explicit in directing how the books and accounts of the county treasurer shall be kept.

COUNTY RECORDER OR REGISTER OF DEEDS.

In a few of the States the office of county recorder or register of deeds is merged with some other county office, in counties where the population falls below a certain amount. A notable example of this is found in both the States of Illinois and Missouri (and there are others), where it is merged with the office of circuit clerk in many counties. The title of the joint office is "circuit clerk and recorder," and the duties of both offices are looked after by one official.

The duties of the county recorder or register of deeds are very similar in the various States, although in some of the Eastern and Southern States the office is called by other names. The usual name, however, is county recorder or register of deeds. In Illinois, Indiana, Iowa, Missouri, Ohio and many other States, it is called "county recorder." In Kansas, Michigan, Minnesota, North Dakota, Wisconsin and many more it is called "register of deeds." In all of the States this office is the repository wherein are kept all records relating to deeds, mortgages, transfers and contracts affecting lands within the county. It is the duty of the recorder or register, as soon as practical after the filing of any instrument in writing in his office entitled to be recorded, to record the same at length, in the order of the time of its reception, in books provided by the county for that purpose; and it is his duty to endorse on all instruments a certificate of the time when the same was filed. All of the States have *some* of the following provisions concerning the duties of the recorder, but these provisions are not common to *all* of the States, viz.: The register or recorder is not allowed to record an instrument of

DIGEST OF THE SYSTEM OF CIVIL GOVERNMENT.

any kind unless it is duly executed according to law; he is not obliged to record any instrument unless his fees are paid in advance; *as a rule*, it is unlawful for him to record any map, plat or subdivision of land situated within any incorporated city, town or village until it is approved by the proper officers of the same. In many States he is forbidden to enter a deed on the records until it has been endorsed "taxes paid" by the proper official; he is required to exhibit, free of charge, all records, and allow copies to be made; he is authorized to administer oaths and take acknowledgments.

CIRCUIT OR DISTRICT CLERK, OR CLERK OF COURT.

In nearly all of the States, each county elects a "clerk of court or courts," sometimes also known as circuit clerk or district clerk, indicating the court with which the office is connected. In some of the States, as has already been stated, the office of clerk of court is merged with some other county office. This is the case in Illinois and Missouri, where in many counties it is connected with the office of county recorder. In Michigan, one official under the name of "clerk" handles the business which usually is given to the clerk of court and county clerk or auditor. In Wisconsin, Missouri, Illinois and other States the name used is "circuit clerk;" in Kansas, Minnesota, North Dakota and many others the office is called "clerk of district court;" while in many of the States, including Indiana, Ohio, Iowa and others, it is called simply "clerk" or "clerk of the court or courts."

The chief duty of this official is to act as clerk of the district or circuit court, and sometimes other courts of inferior jurisdiction. It is the clerk's duty to keep the seals and attend the sessions of their respective courts, preserve all the files and papers thereof, make, keep and preserve complete records of all the proceedings and determinations thereof, and carry out such other duties as may be required by the rules and orders of their respective courts. They must enter of record all judgments, decrees and orders of the court as soon as possible after they are rendered; keep all indictments on file as a public record, have authority to administer oaths, take acknowledgments; take and certify depositions, and are required to exhibit all records free of charge. In nearly all the States the law defines the character of the record books which the clerk of court must keep. Although there is no settled rule in this matter, the general provisions are that he shall keep: First, a general docket or register of actions, in which is entered the title of each action in the order in which they are commenced, and a description of each paper filed in the cause and all proceedings therein; second, a plaintiff's index and defendant's index; third, a judgment book and execution docket, in which he enters the judgment in each action, time of issuing execution, satisfaction, etc., and such other books as the courts or the laws may prescribe.

SHERIFF.

In all of the States the office of sheriff is one of the most important of the county offices. The term of office varies in different States, being usually either two or four years, and in several of the States one party cannot hold the office a second term consecutively. The general provisions outlining the duties pertaining to this office are very much alike in the various States, and the following *resume* of his duties may be said to apply to all of the various States except in a few minor and unimportant details. The sheriff is charged with the duty of keeping and preserving the peace in his county; or, as has been written, "he is the conservator of peace," and it is his duty to keep the same, suppress riots, affrays, fighting, breaches of the peace and prevent crime, and may arrest offenders "on view" and cause them to be brought before the proper magistrate; and to do this, or to execute any writ, warrant, process, order or decree, he may call to his aid when necessary any person or the "power of the county." It is the duty of the sheriff to serve and execute within his county, and return, all writs, warrants, process, orders and decrees of every description that may be legally directed and delivered to him. He is a court officer, and it is his duty to attend, either in person or by deputy, all courts of record held in his county; by virtue of his office he has custody of the jail. It is his duty to pursue and apprehend felons and persons charged with crime and has custody of prisoners. He is not allowed to purchase any property exposed for sale by him as sheriff.

COUNTY SUPERINTENDENT OR COMMISSIONER OF SCHOOLS.

This is an office which exists under one name or another in nearly every State in the Union. The title of the office in a great majority of the States is "county superintendent," but in Michigan, Missouri, Ohio, New York, and possibly one or two other States, the office is termed "school commissioner," and in several of the States the laws provide for a board of county examiners or school commissioners, who are given considerable of the work that in most of the other States is handled by the county superintendent.

The name of this office implies the duties which devolve upon it, and they are very much alike in all of the States. The incumbent of this office is charged with a general supervision over the schools of the county, and must be a fitting person as to education and moral character. As a rule it is their duty to examine and license teachers, but in a few of the States provision is made for a board of examiners. County superintendents are required to visit and inspect the schools at regular intervals, and give such advice and instruction to teachers as may be deemed necessary and proper. They are required to organize and conduct institutes for the instruction of teachers if deemed necessary, and encourage teachers' associations. They introduce to the notice of teachers and the people the best modes of instruction, the most approved plans of building and ventilating school-houses, etc., stimulate school officers to the prompt and proper discharge of their duties. They receive reports from the various school officers, and transmit an abstract of these reports to the State Superintendent, adding a report of the condition of the schools under their charge. In nearly all the States they are forbidden having any interest in the sale of any school furniture, apparatus or books used in the schools. In many States they have authority to annul a teacher's certificate for proper cause, and in general to take such steps and enforce such methods as will elevate and make more efficient the schools under their control.

COUNTY, PROSECUTING OR STATE'S ATTORNEY.

There is a great difference between the various States in the method of handling or attending to the legal business relating to county matters or growing from county affairs. In many of the States the official who attends to this line of work is known as the "county attorney," in other States he is called the State's attorney or prosecuting or district attorney. In a few of the States they divide the State into districts embracing a number of counties, and a district attorney is elected in each district, who in some cases attends to all the legal work of the various counties, and in others he assists the county attorneys in their most important duties and prosecutions. But whatever plan may be followed in the various States, and whatever title may be given to this office, the general duties of the office are very much the same throughout all of the States. It is the duty of the county attorney to commence and prosecute all actions, suits, indictments and prosecutions, civil and criminal, in any court of record in his county in which the "people of the State or county" may be concerned; to prosecute all forfeited bonds and recognizances, and all actions for the recovery of debts, revenues, moneys, fines, etc., accruing to his county; to commence and prosecute all actions and proceedings brought by any county officer in his official capacity; to defend all actions and proceedings brought against his county, or against any county officer in his official capacity; to give legal opinions and advice to the county board or other county officers in relation to their official duties; to attend, if possible, all preliminary examinations of criminals. When requested, he is required to attend sessions of the grand jury, examine witnesses in their presence, give legal advice and see that proper subpœnas and processes are issued; draw up indictments and prosecute the same. The county attorney is required, when requested by the Attorney-General, to appear for the State in cases in his county in which the State is interested. The county attorney makes an annual report to his superior State officer of all criminal cases prosecuted by him.

PROBATE OR COUNTY JUDGE.

The method of handling probate matters is not uniform throughout the various States. In many States the higher courts are given jurisdiction over probate matters, and in others they have created districts in which are held probate courts, whose jurisdiction extends over several counties and takes in other matters besides purely probate affairs. In a majority of the States, however, particularly the Western and Northern States, they elect a county or a probate judge, who holds court and handles the probate matters which arise within his county. The jurisdiction of these county or probate courts is not always confined exclusively to probate affairs, being frequently extended to many other matters, and they generally include such matters as apprenticeship affairs, adoptions, minors, etc. In some of the States they have both a county judge and a probate judge, and in those cases the jurisdiction of the latter is confined to such matters as are in line with probate affairs. In Missouri they have a probate judge, and also a county court, composed of county judges, in whom the corporate powers of the county are vested—as the official county board. In Michigan they have a probate judge and a probate register. The probate judge is generally given original jurisdiction in all matters of probate, settlement of estates of deceased persons, appointment of guardians and conservators, and settlement of their accounts. They take proof of wills, direct the administration of estates, grant and revoke letters testamentary and of administration, appoint and remove guardians, etc.

COUNTY SURVEYOR.

This is an office which is common to nearly all of the States. It is the duty of the county surveyor to execute any survey which may be ordered by any court, or upon application of any individual or corporation, and preserve a record of the surveys made by him. Nearly all of the States provide that certain records shall be kept by the county surveyor, and provide penalties for his failure to place on record the surveys made by him. While he is the official county surveyor, yet the surveys made by him are not conclusive, but may be reviewed by any competent tribunal, and the correctness thereof may be disputed.

COUNTY CORONER.

This is another county office which exists in nearly all of the States. In the average county there is not much work for the coroner, but in the counties in which large cities are located the office is a very important one. In general terms it may be stated that the coroner is required to hold inquests over the bodies of persons supposed to have met with violent or unnatural deaths. In most States he has power to impanel a jury to enquire into the cause of death; but in some of them this is not the case, and he is given power to act alone. He can subpœna witnesses; administer oaths; in certain cases provide for a decent burial, and can bind over to the proper court any person implicated in the killing of the deceased.

OTHER COUNTY OFFICES.

The county offices that have already been mentioned are the principal ones found in all of the States. There are, however, a few other county officials besides those mentioned which exist in many of the States, and which should be briefly mentioned in this connection. These are such offices as county physician, county assessor, county collector, county poor commissioner or superintendent of the county poor-house, master in chancery or court commissioner, county examiners, board of equalization, board of review, etc. The names of these offices imply the duties. These offices do not exist in all of the States, but in nearly every State the law provides for one or more of these county officials.

COUNTY BOARD.

The powers of every county as a body politic and corporate are vested in a county board. This official county board is generally termed the county "board of supervisors," or "board of commissioners," but there are some exceptions to this, like Missouri, where the county board is known as the "county court." There is considerable difference in the make-up of the county board in the various States. In some it is made up of one member from each township in the county. In others the counties are divided into districts, and one member of the county board is chosen from each district. No general description of this could be given that would be accurate, as some of the States follow both of these plans. For instance, in Illinois some of the counties are governed by a board of supervisors, which is made up of one member from each township, while other counties in the same State are governed by a board of county commissioners, consisting of three or more members, each representing districts into which the counties in question are divided.

The general powers of the county board throughout of all the States is about the same, except in minor details. It represents the legislative and corporate powers of the county. One of their number is always chosen as chairman or president, and acts as the presiding officer. The county board has general charge over the affairs of the county. It is their duty to provide county offices, provide desks, stationery, books, fuel, etc.; examine, investigate and adjust claims against the county, and have general care and custody of all the real and personal estate owned by the county. At regular intervals they settle with the county treasurer; examine accounts and vouchers. They locate county roads; determine the amount of county tax, and regularly publish a statement of their proceedings; make statements of receipts, expenditures, etc.; and make all contracts, and do all other acts in relation to the property and concerns of the county necessary to exercise its corporate powers that are not specifically delegated to other county officials.

TOWNSHIP GOVERNMENT.

THE method of township government throughout the different States varies so much that it is impossible in this article to treat of it more than in a general way. In many of the States the townships are not organized as bodies corporate, and in other States in some counties they may have township organization, while in other counties in the same State it does not exist. In cases where there is no township organization the law provides that certain county officials shall attend to the local work, or that work which in other localities is assumed by the township officials. But even where they have township organization the plan of township government in the different States where it exists differs so widely that scarcely any two States may be said to be alike. About the only statements concerning the organized townships that could be made which would apply to all the States are the following: Every organized township in its corporate capacity has power to sue and be sued; to acquire by purchase, gift or devise, and hold property, both real and personal, for the use of its inhabitants, and again to sell and convey the same; and to make all such contracts as may be necessary in the exercise of its powers *as a township*.

In a great many of the States the township government is carried on after a plan very similar to the county and State governments, having various executive officers and a township board in which the corporate and legislative powers of the township are vested. In other States they follow a plan which reserves to the people all corporate and legislative powers, and therefore have no need for a township board, but have various other township officers to carry out the wishes and orders of the voters. Where this plan prevails they hold what is generally termed "town meetings," at which every legal voter of the township has a voice. At these meetings reports are had from the various township officials, and the necessary measures are adopted and directions given for carrying on the township business.

Still other States combine good features from both of the plans above mentioned, and besides the other usual township officials they maintain a township board, which is given certain *restricted* powers, such as those of a review or an auditing board, but they are not vested with the complete corporate and legislative powers of the township, this being reserved in a large measure to the voters, and all questions calling for the exercise of such authority are acted upon at the town meetings. In many of the States the township board just described is made up of three or more of the other township officers, who are ex-officio members of the township board, and they meet at certain times, perform the work required of them, and report to the town meetings.

The principal officials in township organizations in nearly all the States are the following: "Supervisors, or trustees," "clerk," "treasurer," "assessor," "collector," "justices of the peace," "constables," "overseers, supervisors or commissioners of the highways," and "poundmasters," although as has been stated, many of the States do not have all of these officials.

SCHOOL DISTRICT GOVERNMENT.

THE "common school system," or, to speak with greater accuracy the method of governing school districts, in the various States, differs widely, yet all follow in a general way one of two separate and clearly defined methods, being amended in minor respects to meet local conditions and ideas. All of these methods have their excellent points, and yet it has been claimed by eminent educators that no one of them is free from fault and objection, nor has reached perfection. It will be the aim in this article to briefly explain the principal features of the several methods, but it is not possible to go into detail in the matter of giving the system of school government that is followed in each of the many States of the Union. The constitution and statutes of all the States agree, however, upon several points. They aim to provide for a thorough and efficient system of free schools, whereby all the children of the States may receive a thorough common school education; they provide that all lands, moneys and other property donated, granted or received for school, college, seminary or university purposes, and the proceeds thereof, shall be faithfully applied to the objects stated; with two or three exceptions they provide that no appropriation shall be made or public funds applied in aid of any church or sectarian purpose, or to support or sustain any school, academy, seminary, college or university controlled or run in the interest of any church or for a sectarian purpose; and they prohibit the various school officials from holding any interest in the sale, proceeds or profits of any book, apparatus or furniture used in the schools in which they, as officers, are interested.

In many of the States they follow what may be termed the "independent school district" method, inasmuch as each district, so far as its corporate powers are concerned, is entirely separate and independent of other districts. Where this plan is followed the boundaries of each district are clearly defined, and each district is complete within itself. They elect a full set of district officials, and exercise their corporate powers and manage their district affairs within themselves. In this plan the corporate powers of the district are usually vested in a district board, which has general charge of the interests of the district, hires teachers, and makes such contracts and carries into effect such methods as is deemed necessary to raise the grade or aid in the efficiency of the schools. The measure of the authority given to those district boards is not the same in all the States, and in many States it is restricted, and a part of the corporate power is reserved to the people themselves, the officials being required, in all important matters, to carry out the wishes and orders of the people of the district as expressed and decided upon at the "district school meetings."

Another method which is followed in many of the States may be termed the "township system." In such States the law provides for the organization of each township for school purposes, or as one large "district," and each township, so far as its educational interests are concerned, is organized, has the necessary officials and becomes a body politic and corporate. As a general rule, where this method prevails, the townships are divided into three or more sub-districts. All of these sub-districts are a part of the whole, and the finances and general business is generally managed by a township board made up of representatives from each sub-district. This board is generally clothed with the corporate powers, hires teachers, provides fuel and supplies and makes all the contracts necessary to carry on the various schools in the township. As with independent districts, the powers of this board are not alike in all States where the township system prevails, for in some States their power is very much restricted, and is limited to certain official matters, the corporate powers and right to make important contracts being reserved to the people, who decide on these questions at what are termed the school meetings. In a few of the States where they follow the township system they have no official board. This is the case in Indiana, where they elect a township trustee, whose duty it is to look after all the educational interests of the township, subject to the approval of the people at the regular meetings. In most of the States where the township system prevails the law provides for the organization, under certain conditions, of sub-districts into independent districts, which gives them the power to elect their own officers and act independently of the other schools in the township.

In nearly all of the States one of the two general methods given above is followed, with certain changes to make the plan more efficient and satisfactory, and to better meet the desires and needs of the people of the different States. Many of the States combine good features from both these systems, as some of the States have the township system, wherein each sub-district has its own board, and so far as controlling its own affairs is concerned, is independent of all other districts. But local conditions have in many instances made special and local provisions necessary that are different in each State, and while there may be a vast difference in the methods followed, their aim is the same, and, as a whole, the various systems have accomplished the result of giving throughout the length and breadth of the Union the grandest and most efficient system of free schools that the world has ever known.

CITIES AND VILLAGES.

IN all of the States the laws provide for the local government of cities and villages, so that when they attain a certain population they may be separated from, and thus manage their affairs independent of, the township in which they are located, both as to school matters and civil authority. In school affairs provision is made for handling the more complex educational interests of villages and cities—the school boards being made larger, and in many cases the scope of their authority is very much extended. In civil matters provision is made in all of the States for the organization of villages and cities as corporate bodies, separate and distinct from the townships, and providing for the necessary officers to carry on the affairs of the municipality.

General Information on Banking and Business Methods.

Relations between a Bank and its Customers.

[illegible]

Opening an Account.

[illegible]

Deposits.

[illegible]

Discounts, Loans, Etc.

[illegible]

Collections.

[illegible]

Statements and Balances.

[illegible]

Negotiable Paper.

[illegible]

Promissory Notes.

[illegible]

Bills of Exchange.

[illegible]

Checks.

[illegible]

Drafts.

[illegible]

Endorsements.

[illegible]

GENERAL INFORMATION ON BANKING AND BUSINESS METHODS.

Another common form of limiting the endorsement is to enable the payee (when it is made payable to his order) to transfer his title to the instrument without becoming responsible for its payment, and making the party to whom it is transferred assume all responsibility concerning payment. To do this the endorser writes the words "Without Recourse" over his signature, which has the effect of relinquishing his title without making him liable to the holder in case the payor fails to take it up.

Another method of limiting the endorsement is to make it conditional, a good illustration of which is the following: "Pay to John Sims or order upon his delivering to the First National Bank a warranty deed to lot 5, block 4, etc.," below which the endorser places his signature. He can also make it payable to "A. B. only," or in equivalent words, in which case "A. B." cannot endorse it over.

In fact, the endorser has the power to limit his endorsement as he sees fit, and either to lessen or increase his liability, such as either "waiving notice of demand;" making his endorsement a "general and special guaranty of payment" to all future holders, etc., but he cannot, by his endorsement, either increase or lessen the liability of any other endorser on the instrument.

An endorser, as a rule, is entitled to immediate notice in case the paper fails to pay it. This is the case in nearly all of the United States, as it has been a rule of the "law merchant" for many years. A few modifications, however, of the general "law merchant" have been made by statute in several of the States, relating to negotiable paper, in changing the endorser's liability by rendering his contract absolute instead of conditional, making notice unnecessary unless he suffers damage through want of it, or requiring a judgment to be first recovered before he can be held. In the absence, however, of statutory provisions of this kind, and they only exist in a few of the States, it may be said that to hold endorsers they must have prompt notice of non-payment, and it may be said to be a general rule of the "law merchant" that all parties to negotiable paper as endorsers who are entitled to notice are discharged by want of notice. The demand, notice and protest must be made according to the laws of the place where payable.

The term *Protest* is applied to the official act by an authorized person (usually a Notary Public), whereby he affirms in a formal or prescribed manner in writing that a certain bill, draft, check or other negotiable paper has been presented for acceptance or payment, as the case may be, and been refused. This, and the notice of the "Protest," which must be sent to all endorsers and parties to the paper is to notify them officially of its failure.

GUARANTY.

A "GUARANTOR" is one who is bound to another for the fulfillment of a promise, or of an engagement, made by a third party. This kind of contract is very common. According to the "statute of frauds" it must be in writing, and unless it is a sealed instrument there must be a consideration to support it. As a rule it is not negotiable, so as to be enforced by the transferee as if it had been given to him by the guarantor, but this depends upon the wording, as, if it contains all the characteristics of a note, payable to order or bearer, it will be held negotiable. A contract of guaranty is construed strictly, and, if the liability of the principal be materially varied by the act of the party guaranteed, without the consent of the guarantor, the guarantor is discharged. The guarantor is also discharged if the liability or obligation is renewed, or extended by law or otherwise, unless he in writing renews the contract. In the case of a bank incorporated for twenty years, which was renewed for ten years more without change of officers, the courts held that the original sureties could not be held after the first term.

The guaranty can be enforced even though the original debt cannot, as is the case in becoming surety for the debt of a minor. A guarantor who pays the debt of the principal is entitled to demand from the creditor all the securities he holds, or of the note or bond which declares the debt; and, in some States, the creditor cannot fall back upon the guarantor until he has collected as much as possible from these securities and exhausted legal remedies against the principal. If the debt or obligation be first incurred and completed before the guaranty is given, there must be a new consideration or the guaranty is void.

A guaranty is not binding unless the guarantor has notice of its acceptance, but the law presumes this acceptance when the offer of guaranty and acts of the party to whom it is given, such as delivery of goods or extending credit are simultaneous. But an offer to guarantee a future operation does not bind the offerer unless he has such notice of the acceptance as will afford him reasonable opportunity to make himself safe. A creditor may give his debtor some indulgence or accommodation without discharging the guarantor, unless it should have the effect of prejudicing the interests of the guarantor, in which case he would be released. Generally a guarantor may, at any time, pay a debt and so, at once, have the right to proceed against the debtor. Where there has been failure on the part of the principal and the guarantor is looked to, he must have reasonable notice—and notice is deemed reasonable if it prevents the guarantor from suffering from the delay.

It is, in many cases, difficult to say—and upon it rests the question of legal liability—whether the promise of one to pay for goods delivered to another is an original promise, as to pay for one's own goods, in which case it need not be in writing; or a promise to pay the debt of guaranty the promise of him to whom the goods are delivered, in which case it must be in writing. The question generally resolves itself into this: To whom did the seller give and was authorized to give credit? This is a question of fact and not of law. If the books of seller show that he charged them to the party to whom he delivered them, it is almost impossible for him to hold the other party for it, but if on the other hand it is shown that he regarded the goods as being sold to the party whom it is desired to hold, but delivered them to another party and it is so shown on his books, it is not regarded as a guaranty, but an original or collateral promise, and would make the party liable. In general, a guarantor of a bill or note is not entitled to such strict and exact notice as an endorser is entitled to, but only such notice as shall save him from actual loss, as he can not make the want of notice his defense unless he can show that it was unreasonably withheld and that he suffered thereby. There is a marked difference in the effect of a guaranty of the "payment," or of the "collection" of a debt. In the first case, the creditor can look to the guarantor at any time; in the latter, the creditor must exhaust his legal remedies for collecting it.

ACCOMMODATION PAPER.

AN accommodation bill or note is one for which the acceptor or maker has received no consideration, but has lent his name and credit to accommodate the drawer, payee or holder. He is bound to all other parties just as completely as if there were a good consideration, for, if this was not the case, it would be of no value to the party accommodated. He is not allowed to set up want of consideration as a defense as against any holder for value. But he is not bound to the party whom he thus accommodates, no matter how the instrument may be drawn.

IDENTIFICATION.

THE mere act of identifying a party or making him known to a banker carries with it no liability on the part of the party who thus performs it, unless it can be shown there was fraud or collusion. Customers of banks are frequently asked to identify and make known to their own bankers, strangers who desire checks or drafts cashed or other accommodations. In some cases a mere introduction is all that is necessary, but only because the banker relies upon the honor and integrity of his customer, knowing that an improper person would not be introduced, for in a case of this kind the bank assumes all the risk. Generally speaking, however, it is an almost invariable rule with bankers, as it should be, to require their customer to endorse all drafts or checks which are honored for the stranger. In this case the endorser becomes personally liable to the bank if any or all of the drafts or checks prove worthless.

An endorsement which is frequently made by parties who are asked to identify others is to merely indicate that they know the party to be the payee named in the check or that the signature of the payee or party is correct. This is done by writing the words "Signature O. K." under the party's name and signing it. This has the effect of guaranteeing that the party's name is so written and that it is his proper signature. It does not guarantee that the check or draft is good or will be paid, but merely an expression that the signature is correct, and the only liability assumed is that he will pay the amount in case the signature proves a forgery. Many banks, however, will not accept paper endorsed this way and justly so, for it throws upon them the burden of the risk.

RECEIPTS AND RELEASES.

ANY acknowledgment that a sum of money has been paid is a receipt. A receipt which reads "in full" though admitted to be strong evidence is by no means legally conclusive. If the party signing it can show an error or mistake, it will be admitted in his favor. Receipts for money will be held open to examination, and the party holding it must abide the results of such examination, the great aim of the law being to administer strict justice. A receipt may be of different degrees of explicitness, as the word "Paid" or "Received Payment" written on a bill. A "release" is simply a form of receipt, but is more binding upon the parties, inasmuch as, if properly drawn, under seal, for a consideration, it is a complete defense to any action based on the debts or claims so released. Herein, releases differ from receipts. A release is in the nature of a written contract and therefore cannot be controlled or contradicted by evidence, unless on the ground of fraud. But if the words are ambiguous, or may have either of two or more meanings, evidence is receivable to determine the meaning.

INFANTS AND MINORS.

THE incapacity of a person to make a valid contract may arise from several causes, and the fact of being an infant, or minor, is one of them. The general rule of law may be stated as being that the contract of an infant or minor is not always void, but is voidable, and in many cases special exception is made, giving validity to their contracts for necessaries. By being voidable, but not void in themselves, means that the infant has the right to disavow and annul the contract, either before or within a reasonable time after he reaches his majority. He may do this by word only, but a mere acknowledgment that the debt exists is not enough, and it must be substantially a new promise.

AGENCY.

THERE are a few well-settled and important rules of law governing the matter of agents and agency, which every business man should understand thoroughly. The relation of principal and agent implies that the principal acts by and through the agent. A principal is responsible for the acts of the agent *only* when he has actually given *full* authority to the agent, or when he has by his words, or his acts, or both, caused or permitted the person with whom the agent deals to believe him clothed with this authority. This is a point which is not always thoroughly understood, but it is a well-settled principle of law. There are two kinds of agents—general and special. A general agent is one authorized to represent his principal in all his business, or in all his business of a particular kind, and his power is limited by the usual scope and character of the business he is empowered to transact. If he is given out as the general agent, the principal is bound, even if the agent transcends his actual authority, but does not go beyond the natural and usual scope of the business.

On the other hand, a special agent is one authorized to do only a specific thing, or a few specified things, or a specified line of work. If this special agent exceeds his authority, it may be stated as an almost invariable rule that the principal is not bound, because the party dealing with the agent must inquire for himself and at his own peril, into the extent and limits of the authority given to the agent. Especially is this the case where the party knew that the agent had been or was engaged in attending to a particular and specified line of work connected with the business of the principal. The party, however, is not bound by any special reservations or limitations made secretly by the principal of which he had no reasonable or easy means of having notice. The authority of an agent may be given by the principal, by writing or orally, or may be implied from certain acts. Thus if a person puts his goods into the custody of another whose business it is to sell such goods, he authorizes the whole world to believe that this person has them for sale; and any person buying them honestly, in this belief, would hold them. If one, knowing that another had acted as his agent, does not disavow the authority as soon as he conveniently can, but lies by and permits a person to go on and deal with the supposed agent, or lose an opportunity of indemnifying himself, this is an adoption and confirmation of the acts of the agent.

A principal is bound by the acts of an agent even after the revocation of his agency, if such revocation has not been made public or is unknown to the party dealing with the agent. An agent can generally be held personally liable if he transcends his authority; but this is not the case if the party with whom he dealt knew that the authority was transcended.

ORIGIN AND HISTORY OF BANKING.

IN general, banks may be said to be credit institutions or dealers in credit. John Jay Knox once said that "the exchanges of the modern world are barter, effected by the indirect agency of the credit system, and banks and bankers are the machinery by which this is done." Metallic money and its representative, the circulating note, are only the small change of "Trade" employed in the settlement of balances and small purchases and payments. This fact is illustrated by the operations of the New York clearing house. The exchanges have been about [illegible] millions of dollars during the past thirty years while the balances paid in money have only been about [illegible] millions, or about four per cent. of the amount of the settlements.

It has always been claimed that the business of banking originated with the Venetian money changers who displayed their wares and moneys on the streets and thus supplied those in need of change. According to the most eminent authorities the earliest banking institution in Europe was the Bank of Venice, which was founded in 1171, and was based upon a forced loan of the government. Funds deposited in it could be transferred to others on the books of the bank at the pleasure of the owner, but they could not be withdrawn. The perpetual annuities of the British debt are handled in a very similar manner at the present day. The Bank of Venice was continued until 1797. In 1401, the Bank of Barcelona was formed. At a period much earlier than this, the Jewish money-dealers had invented what are known as "foreign bills of exchange," but it is said that this bank was the first institution that made a business of negotiating and handling them. The Bank of Genoa commenced operation in 1407 and for centuries was one of the principal banks of Europe. It was the first to issue circulating notes—which were passed only by endorsement, not being payable to bearer.

The Bank of Hamburg, established in 1619, was a bank of both deposit and circulation based on fine silver bars. This bank, like nearly all of that early time, had, as a principal object, the protection of the people from worn, sweated, clipped and plugged coins, or coins of certain empires that were reduced in standard value. The remedy generally adopted was to lock up the debased and depreciated coins and circulate the credit granted for them. Various other banks sprang into existence throughout Europe, many of them being powerful government agencies, and in many cases exerted a wide influence in shaping the destinies of empires.

In 1694 the Bank of England was established, and there is no banking institution in the world equal to it in the management of national finances. The Bank of France was authorized in 1803. It is not a fiscal agent of the government as is that of England. It does not collect or disburse the revenues of the exchequer but it lends to it largely, while its credits, in the form of circulating notes and other acceptances, have borne the government safely through extraordinary needs.

It is claimed that the first organized bank in the United States had its origin in the formation of a banking company without charter June 17th, 1780, by the citizens of Philadelphia, and first action by Congress was taken June 21, of the same year in reference to this proposed association. Two years afterward, a "perpetual charter" was granted to the Bank of North America at Philadelphia. In 1784 the State of Massachusetts incorporated the Massachusetts Bank. The Bank of New York was chartered in March, 1791, although it had been doing business since 1784, under articles of association drawn by Alexander Hamilton. Most of these institutions are still running and have been converted into national banks. The Bank of the United States was organized in 1791. The most of the stock was owned by the United States Government, but later the Government interest was disposed of, and in 1841 the bank failed.

State banks were organized rapidly, and private banking firms sprang into existence and the business of banking assumed immense proportions.

In 1863, the NATIONAL BANK SYSTEM was adopted and in 1864 the National Bank Bureau of the Treasury Department was organized, the chief officer of which is the comptroller of the currency. In March, 1865, an act was passed providing for a ten per cent. tax on notes of any person or State bank issued for circulation, and making an exception of National banks. This had the effect of taxing the State bank circulation out of existence. As the National banking system has proved one of the most efficient and satisfactory methods the world has ever known, it will be of interest to review here some of its principal features. Under this act National banks may be organized by any number of persons not less than five. Not less than one-third of the capital must be invested in United States bonds, upon which, circulating notes may be issued equal to 90 per cent. of the par value of the bonds. These circulating notes are receivable at par in the United States in all payments except for duties on imports, interest on the public debt and in redemption of the national currency. The National banks are required to keep a certain reserve; they are authorized to loan money at the rate of interest allowed by the various States—when no rate is fixed by the laws of the State, the banks may charge 7 per cent. Shareholders are held individually liable, equally and ratably, for all debts of the association to the extent of the amount of their stock, in addition to the amount invested therein. The banks are required, before the declaration of a dividend, to carry one-tenth part of their net profits of the preceding half year to a surplus fund until the same shall amount to 20 per cent. of the capital; and losses and bad debts must be deducted from net profits before any dividend is declared. A receiver may be appointed by the comptroller to close up under his supervision the affairs of any national bank which shall fail to keep good its lawful money reserve or which may become insolvent. While there have been national bank failures, there has never been any loss to the public whatever on the circulation. A suit may be brought for forfeiture of the charter of a bank if the directors shall knowingly violate the law; and in such cases they may be held liable in their individual capacity. There are other restrictions in the law—such as for instance, the prohibition against loaning to any one borrower of more than ten per cent. of the capital; or the holding of any real estate except such as is required for banking purposes, or the granting of loans upon the security of the bank stock.

The national bank circulation has been gradually growing less during the past ten years, as the United States bonds available are quoted so high above par and the rate of interest so low that there is but little profit to the banks in it. All of the States have laws regulating State Banks and providing certain restrictions, but as the laws of the various States are not alike it is impossible to give a general description of the matter that would apply to all the States. The laws, however, provide for and require State banks to hold a certain reserve, and at regular intervals they make full statements as to their condition and their affairs are examined into by certain State officials at frequent intervals. The laws of all the States have reached a high degree of perfection in the method of regulating and overseeing State banks, and the almost universal soundness and reliability of these institutions reflect credit upon the laws under which they exist.

CLEARING HOUSE.

THE Clearing-House is the place where the exchanges of the banks are made in all the principal cities of the world. The clearing-house system was first established in London about the beginning of the present century. It was first introduced into this country by the banks of the city of New York organizing an association, under the name of the New York Clearing House, which commenced operations Oct. 11, 1853. At that time it consisted of fifty-two banks, but five of them were soon closed because of their inability to meet its requirements. Clearing Houses have since been established in nearly all of the principal cities of the continent.

In all cities a bank receives large amounts of bills of and checks on other banks, so that at the close of each day's business every bank has, in its drawers, various sums thus due it by other banks. It is, in like manner, itself the debtor of other banks, which have during the day received its bills and checks drawn upon it. Prior to the establishment of the clearing-house it was necessary for each bank, every morning, to make up its account with every other bank, and to send its porter or agent to present the bills and checks so received to the debtor banks for payment. The balances were adjusted by payments in gold, which became so laborious, dangerous, and complicated, that the balances were settled only weekly instead of daily—a plan that resulted in great risk and evil. This was obviated by the clearing-house system, through which the settlements are so simultaneously and quickly effected that in New York the transactions in one single day have amounted to over $[illegible], in adjusting which the exchanges were settled in the space of an hour. Besides saving a vast amount of work, book-keeping and expense, it enabled the banks by united aid to strengthen each other in times of excitement and financial panic.

The following is the manner in which the settlements are made in about all the clearing-houses of this country: The clearing-room is provided with a continuous line of desks, one for each bank that is a member of the association, each desk bearing the name and number of the bank. Each bank is represented every morning, at the hour fixed for settlement, by two clerks, one a messenger who brings with him the checks, drafts, etc., that his bank has received during the day previous upon the other banks called the "exchanges," and these are assorted for each bank and placed in envelopes. On the outside of each envelope is a slip on which are listed the amounts of the various items which it contains. The messengers take their places in a line outside the row of desks, each opposite the desk assigned to his bank, while at each desk is a clerk with a sheet containing the names of all the banks in the same order as the desks, with the aggregate amounts which his bank's messenger has against each bank. Just previous to the hour fixed for making the exchanges the manager takes his position and calls the house to order. At a signal the bell rings and each messenger moves forward to the desk next his own and delivers the envelope containing the checks, etc., for the bank represented at that desk to the clerk at that desk, together with a printed list of the banks in the same order, with the amount opposite each bank. The clerk receiving it, signs and returns it to the messenger, who immediately passes on to the next desk; then to the next, and so on until he has made a complete circuit and has again reached the desk of his own bank—the starting point. All the other messengers moving in the same manner; each messenger has, by this means, visited every bank and delivered to each everything his bank held for it, taking a receipt for the same; and at the same time each bank has received all the exchanges that every other bank had against it. This operation even in the greatest clearing houses only consumes from ten to fifteen minutes.

This enables the banks to know at once the exact balance for or against it, as the clerks immediately enter from the slips on their own sheets the aggregate amount from each bank, and the difference between the total amount brought by them, which at once shows the balance due to or from the clearing house to each bank.

This is reported to their banks, and the balance is paid to or drawn from the clearing house, thus at once settling the accounts between all the banks. The lists are "proved" carefully, and certain fines are laid for all errors, tardiness, etc.

CHRONOLOGICAL ARRANGEMENT

—OF—

ANCIENT, MEDIEVAL AND MODERN HISTORY

The chief aim of this Chronological History is to give in a comprehensive and attractive form the principal events of the history of the world free from unnecessary details. For convenience this history is arranged under—I. Ancient History. II. Medieval History. III. Modern History. The latter is given—First. From the beginning of the Sixteenth Century to American Revolution. Second. From the birth of the United States to the present time by countries.

Ancient History

B. C.

[illegible]

ANCIENT, MEDIEVAL AND MODERN HISTORY.

[illegible]

ANCIENT, MEDIEVAL AND MODERN HISTORY.

A. D.

[illegible] Augustine made Bishop of Hippo (died [illegible]).
Alaric in Greece.
Stilicho attains chief power under Honorius.
[illegible] The Britons ask aid of Honorius against the Picts and Scots.
[illegible] Deaths of Martin of Tours and Ambrose of Milan.
[illegible] Chrysostom, Bishop of Constantinople (died [illegible]).
[illegible] Alaric ravages Italy.
[illegible] Battle of Pollentia.
Defeat of Alaric by Stilicho.
[illegible] The Vandals, Alans and Suevi invade Spain.
[illegible] The Roman legions recalled from Britain; final withdrawal about [illegible].
410 Sack of Rome by Alaric.
Death of Alaric.
Pelagius begins to preach about this time.
412 Proclus the philosopher born (died [illegible]).
414 Marriage of Ataulphus, King of the Goths, to Placidia, daughter of Theodosius the Great.
Persecution of the Christians in Persia begins; lasts thirty years.
420 Death of St. Jerome.
Orosius, the Spanish presbyter and historian, flourished.
[illegible] Death of Honorius at Ravenna.
[illegible] Administration of Ætius begins, lasting about thirty years.
The Traveler's Song published.
[illegible] Nestorius, Patriarch of Constantinople, banished [illegible].
[illegible] The Vandals under Genseric invade Africa.
Death of Theodore, Bishop of Mopsuestia.

[illegible]

Medieval History

[illegible]

ANCIENT, MEDIEVAL AND MODERN HISTORY.

A. D.

[illegible]

Modern History.

[illegible]

ANCIENT, MEDIEVAL AND MODERN HISTORY.

1648 Canadians at war with the Indians.
The House of Brandenburg acquires Halberstadt and Minden.
New Amsterdam contains about 1,000 inhabitants.
1649 Trial and execution of Charles I.
Massacre and capture of Drogheda, Ireland, by Cromwell.
Confession of Faith.
1650 Marquis of Montrose beheaded in Scotland.
1651 Leopold I. made King of Hungary.
Charles II. crowned at Scone, Scotland, Jan. 1.
Battle of Worcester, Sept. 3, and defeat of royalists.
Charles II. flees to France.
"Barebones" Parliament.
Birth of Fenelon, died 1715.
English Navigation Act.
1652 England at war with Holland.
The Dutch, under Van Tromp, "sweep the Channel."
De Ruyter defeated by Blake.
1653 Negro insurrection suppressed in Mexico.
Peace between England and Holland.
Death of Van Tromp.
Long Parliament dissolved by Cromwell, April 20. He becomes Lord Protector, Dec. 16.
1654 Jesuits establish themselves among the Onondaga Iroquois.

[illegible]

Modern History.

From A. D. 1763 to the present time, by Countries.

CHINA.

[illegible]

INDIA.

[illegible]

ANCIENT, MEDIEVAL AND MODERN HISTORY.

[illegible]

RUSSIA.

[illegible]

TURKEY.

[illegible]

GREECE.

[illegible]

ANCIENT, MEDIEVAL AND MODERN HISTORY.

French and Russian fleets defeat the Turks and Egyptians.
Independence of Greece established.
1828 The Turks evacuate the Morea.
1829 Turkey surrenders Missolonghi.
Treaty of Adrianople.
1831 President D'Istria assassinated.
1833 Accession of Otho I.
1843 Insurrection in Athens; National Assembly; new constitution adopted.
1850 Piræus blockaded by a British fleet.
England demands indemnity for injury to British subjects.
French intervention sought.
Greece forced to yield.
1854 Revolt of Albanians.
English and French occupy Greece.
Neutrality in Russo-Turkish war declared.
1857 Greece evacuated by the French and English.
1862 Serious insurrections in Greece.
Otho I. forced to leave Greece.
Prince Alfred, of England, declared King.
Austria declares for Otho I.
1863 National Assembly declares Alfred elected King.
England refuses to allow his accession.
Prince William, of Denmark, elected King, March 18, and becomes King George I., Nov. 2, 1863; new Constitution adopted.
1867 King George I. married to Princess Olga, of Russia.
1870 Trouble with the brigands, who kill many English prisoners.
1875 Neutrality observed in Herzegovinian insurrection.
1876 Declares for neutrality in Servian war.
1878 Thessalians aided by Greeks against the Turks.
1880 Berlin conference considers question of Greek and Turkish frontiers.
1881 Convention with Turkey, July 2.
Thessaly ceded to Greece.
1884 Serious fire at royal palace, Athens, Aug. 4.
1889 Princess Sophie of Russia and the Crown Prince married, October 27.
1890 Greek Ministry resigns, October 23.
1891 Prof. Waldstein discovers rare jewels in the ruins of Eretria, March.
1893 Ministry resigned May 10, and succeeded by a new Cabinet, with M. Tricoupis as premier, Nov. 11.

ITALY.

1775 Death of Pope Clement XIV. and elevation of Pio VI.
1796-'97 Bonaparte's first victories in Italy.
1797 Treaty of Campo Formio.
France and Austria divide the Venetian States.
The Cis-Alpine republic founded.
1798 Second invasion of the French.
Pope Pius VI. deposed by Bonaparte.
1799 Defeat of the French at Trebia, by the Russians, under Suwarrow.
1800 Death of Pio VI.; Pio VII. Pope.
Bonaparte crosses the Alps.
Battle of Marengo, June 14; total defeat of Austrians.
1802 The Cis-Alpine republic remodeled as the Italian republic; Bonaparte President.
1805 Napoleon crowned King of Italy, May 26.
Eugene Beauharnois made Viceroy of Italy.
1806 The Treaty of Presburg deprives Austria of her Italian possessions.
1814 Downfall of Napoleon.
Overthrow of the Kingdom of Italy.
1815 Establishment of the Lombardo-Venetian Kingdom for Austria.
Genoa added to the Sardinian crown.
1823 Death of Pope Pio VII.; Leo XII. becomes Pope.
1829 Death of Leo XII.; Pio VIII. becomes Pope.
1831 Death of Pope Pio VIII., and elevation of Gregorio XVI.
Death of Carlo Felix, and extinguishment of the direct male line of the House of Savoy.
The crown falls to Prince Carlo Alberto.
The "Young State Party" formed by Mazzini.
Insurrection in Central Italy.
1837 King Charles Albert, of Sardinia promulgates a new Code.
1846 Death of Pope Gregorio XVI.; Pius IX. becomes Pope.
1848 The King of Sardinia grants a Constitution and openly espouses the cause of Italian regeneration against Austria.
Insurrection in Lombardy and Venice against Austrian power; revolt is supported by the King of Sardinia.
The Pope supports the movement for Italian independence, June.
War between Sardinia and Austria.
Lombardy annexed to Sardinia, June 29.
Revolution at Rome; flight of the Pope to Gaeta.
1849 The Sardinians, after repeated reverses, are totally defeated by the Austrians at Novara, March 23.
Close of the war, and recovery of Lombardy by Austria.
Carlo Alberto abdicates in favor of his son, Victor Emmanuel II., March 23; dies July 28.
The Roman republic formed.
Rome captured by the French army, under Marshal Oudinot.
The republic overthrown, and the Pope restored.
1850 Ecclesiastical jurisdictions abolished in Sardinia.
Arrest of the Archbishop of Turin.
1852 Count Cavour Minister of Foreign Affairs.
1853 Revolt in Milan subdued.
1855 Sardinia joins the alliance of France, England and Turkey against Russia, and takes part in the Crimean war.
1856 Unsuccessful revolt in Sicily.
1857 Diplomatic rupture between Sardinia and Austria.
1859 Quarrel between Sardinia and Austria, caused by former power refusing to disarm.
France espouses the cause of Sardinia, and sends an army to her assistance.
The Austrians cross the Ticino, April 27.
The French army reaches Genoa, May 2.
Battles of Montebello, May 20; Palestro, May 30, 31; Magenta, June 4; Malegnano, June 8; Solferino, June 24.
Total defeat of Austrians.
Revolutions in Tuscany, Parma, Modena, Bologna, Ferrara, etc.
Peace of Villafranca, July 11.
Western Lombardy annexed to Sardinia.
Protest of Tuscany, and declaration for a United Kingdom.
The people incited to arms by Garibaldi.
The Pope appeals to Europe against the King of Sardinia, July 12.
The Italian Duchies declare in favor of annexation to Sardinia.
New constitution for Sardinia.
Alliance between Tuscany, Modena, Parma and the Romagna formed, Oct. 10.
Peace of Zurich, Nov. 10; part of the Papal States and the Duchies of Parma and Modena ceded to Sardinia.
The Emperor Napoleon advises the Pope to give up his revolted States, Dec. 31.
1860 The Pope refuses the Emperor's proposal and denounces him, Jan. 8.
A new ministry formed by Cavour, Jan. 18.
Tuscany, Parma, Modena and the Romagna vote for annexation to Sardinia, March 9.
Savoy and Nice ceded to France by Sardinia.
The French troops leave Italy in May.
Garibaldi lands in Sicily, May 11.
Declares himself Dictator, and drives the Neapolitans from Sicily in the battles of Calatafimi and Melazzo, July 20.
He invades Naples with his little army, Sept. 7.
Insurrection in the Papal States in September. Sardinian army enters them, and defeats the Papal troops, Sept. 18, and takes Ancona, Sept. 29.
The Sardinian army, under the King, enters the Neapolitan territory; defeats the Neapolitans, at Isernia, Oct. 17.
Garibaldi defeats the Neapolitans, at the Volturno, Oct. 1.
Meets Victor Emmanuel, Oct. 26, and salutes him as "King of Italy."
Sicily and Naples vote for annexation to Sardinia, Oct. 21.
Victor Emmanuel enters Naples as King, Nov. 7.
Garibaldi resigns the Dictatorship and retires to Caprera.
1861 The first Italian Parliament assembles, Feb. 18.
Parliament decrees Victor Emmanuel "King of Italy," Feb. 26.
The new kingdom recognized by England, March 31.
The Pope protests against the new kingdom, April 15.
Death of Cavour, June 6.
Unsuccessful revolt in Calabria, by Jose Borges, in the interest of Francis II.
1862 Ratazzi forms a new ministry.
Naples declared in a state of siege.
Ratazzi's ministry overthrown and a new one formed by Farina.
Garibaldi endeavors to wrest Rome from the Pope.
He is made prisoner at Aspromonte, by the Italian army.
1863 Commercial treaties with France and Great Britain.
1864 Treaty with France for the evacuation of Rome by the French in February, 1867.
Transfer of the Capital from Turin to Florence.
1865 Bank of Italy established.
New Parliament meets at Florence.
The insurrections at Turin suppressed.
Brigands cause much trouble.
1866 The Austro-Italian war begins.
Alliance with Prussia.
Italy declares war against Austria, June 20.
Italians cross the Mincio, June 23.
Battle of Custozza, June 24, and defeat of the Italians by the Archduke Albrecht.
Battle of Lissa.
Defeat of the Italian fleet, July 20.
Peace of Prague, Aug. 23; Eastern Lombardy and Venetia added to the Kingdom.
Treaty of Nicholsburg, Aug. 26; close of the war.
Cession of Venetia to the Italian kingdom.
King Victor Emmanuel enters Venice, Nov. 7.
1867 Insurrection in the Papal States.
Garibaldi placed under arrest.
The French enter Rome.
Garibaldi defeated at Mentana.
1868 Railway over Mont Cenis opened.
Crown Prince Humbert marries Princess Margherita.
1869 Ecumenical Council held at Rome.
Severe earthquake at Florence.
1870 Dogma of Infallibility proclaimed by the Council.
Arrest of Mazzini at Palermo.
The Papal States entered by the Italian army, and Rome occupied, Sept. 20.
Papal States a part of the Kingdom of Italy, Oct. 9.
Pope Pius IX. issues bull of excommunication against the government, Nov. 1.
Rome evacuated by the French, Aug. 17.
Revolution in Rome imminent.
The Pope takes refuge in the castle of St. Angelo.
Rome annexed to Italy, and made the Capital of the kingdom by royal decree, Oct. 9.
The Italian Duke of Aosta elected King of Spain.
1871 The government transferred from Florence to Rome, July.
Opening of the Mt. Cenis Tunnel.
1872 Death of Mazzini.
Great eruption of Mount Vesuvius. Serious inundations throughout the peninsula.
1873 Suppression of the convents at Rome.
Expulsion of Jesuits from Italy.
1874 General assembly of free Christian churches in Italy.
Brigands cause great trouble.
The government suppresses the Camorra.
1875 Visit of the Emperors of Austria and Germany to the King of Italy.
Garibaldi takes oath of allegiance to the government, and becomes a member of the Chamber of Deputies.
Ratification of a treaty of commerce with Great Britain.
Six new cardinals appointed.
1876 Italy anti-Turkish in the eastern question.
Attempted assassination of King Humbert, Nov. 7.
1877 The celebrated "Antonelli" case dismissed.
1878 Death of Victor Emmanuel, Jan. 9.
Attempted assassination of King Humbert I., Nov. 17.
Death of Pope Pius IX., Feb. 7.
Leo XIII. elected Pope, Feb. 20.
1880 Elections favorable to the ministry of Cairoli.
The monster ironclad Italia successfully launched.
Resignation of Garibaldi as Deputy, and retirement to Rome.
1881 Cairoli ministry overthrown and a new one founded by Depretis.
Reform Bill passed by the Senate, Dec. 21.
1882 Electoral law passed.
Death of Garibaldi, June 2.
1883 Discovery of site of the celebrated Atrium, at Rome, Nov. 8.
1884 The cholera rages in Naples.
1889 Statue of Bruno unveiled at Rome, June 9.
1890 Statue of Victor Emmanuel unveiled, Sept. 20.
1891 Crispi Resigns the Premiership and Rudini appointed, Feb. 9.
Baron Fava, Minister to the United States, recalled, March 31.
1893 Pope Leo XIII. celebrates his 83d birthday.
King Humbert and Queen Margaret celebrate their silver wedding.

SPAIN.

1767 Jesuits expelled from the kingdom.
1771 Falkland Islands ceded to England.
1775 War with Portugal resumed.
1779 War with England renewed.
France and Spain besiege Gibraltar.
1783 England cedes Balearic Isles to Spain at peace of Versailles.
1794 French invade Spain.
1796 War again with England.
1797 Battle of Cape St. Vincent; defeat of the Spanish fleet, Feb. 14.
1800 Spain cedes Parma to France.
1801 Treaty with Portugal at Badajos.
Treaty of Madrid with France.
1802 Treaty with England at Amiens.
1804 Renewed war with England.
1805 Battle of Trafalgar, Oct. 21; total defeat of French and Spanish fleets by English, under Nelson.
1807 Invasion of Spain by the French.
Treaty of Fontainebleau.
1808 Territory demanded by France.
Spanish fortress seized.
The French take Madrid.
Charles IV. abdicates in favor of Napoleon, May 1.
Massacre of 300 French in Madrid, May 2.
Napoleon assembles the notables at Bayonne, May 25.
Ferdinand VII. abdicates.
Napoleon I. gives crown to his brother Joseph Bonaparte, who enters Madrid, July 12, but is driven out, July 29.
The French defeated at Vimiera, Aug. 21, by the English.
Battle of Logrono; defeat of the patriots.
Battle of Durango; the French victorious.
The French retake Madrid, and restore King Joseph Bonaparte, Dec. 2.
Napoleon enters Madrid, Dec. 4.
1809 Battle of Corunna and death of Moore, Jan. 16.
Surrender of Saragossa.
Spain entered by Sir Arthur Wellesley, who crosses the Douro.
Defeat of the French at Talavera, July 28.
Spanish defeated at Ocana, Nov. 19.
Severe battle of Molinos del Rey, Dec. 21.
1810 Granada, Seville and Astorga seized by the French.
Capture of Ciudad-Rodrigo by Marshal Ney, July 10.
1811 Wellington defeats the French at Fuentes d'Onoro, May 5, and at Albuera, May 16.
Tarragona taken by Suchet.
King Joseph returns to Madrid.
Spanish defeated by Soult at Lorca.
1812 Wellington victorious at Ciudad-Rodrigo, Jan. 19.
Badajos stormed and carried, April 6.
Defeat of the French at Salamanca, July 22.
1813 English, under Wellington, occupy Madrid.
English successful at Castalla, April 13; Vittoria, June 21, and Pyrenees, July 28.
The French driven out of Spain. Wellington crossing the Bidassoa and follows them into France.
1814 Ferdinand VII. restored.
1817 The slave trade abolished for a compensation.
1820 Revolution under Nunez del Riego begins in January.
Ferdinand swears to the constitution of the Cortes.
1823 The Cortes remove the king to Seville, and thence to Cadiz, March.
Intervention of France in behalf of the king.
French army enters Spain, April 7.
Cadiz invested, June 24.
Battle of the Trocadero, Aug. 31.
Rebels defeated and the revolution crushed.
The king again restored.
Execution of Riego and the patriot leaders.
1828 The French evacuate Cadiz.
1829 Cadiz proclaimed a free port.
1830 The Salique law abolished.
1833 Death of Ferdinand VII.; his queen assumes the government as Regent during the minority of her daughter, Isabella II.
Don Carlos claims the throne.
1834 The Quadruple Treaty of France, England, Spain and Portugal guarantees the right of Queen Isabella to the throne.
Don Carlos enters Spain and claims the crown.
Beginning of the Carlist war.
1836 Defeat of Carlists at battle of Bilbao.
1837 Dissolution of monasteries.
1839 Success of the government forces.
Don Carlos takes refuge in England.
1840 Espartero, commander of the royal forces, becomes the real ruler of Spain.
The Queen Regent Christina abdicates and leaves Spain.
Espartero expels the Papal Nuncio.
1841 Espartero declared, by the Cortes, Regent during the young Queen's minority.
Insurrection in favor of Christina quelled.
1842 Insurrection at Barcelona against Espartero; he bombards the city, Dec. 3, and receives its surrender, Dec. 4.
1843 Uprising against Espartero at Barcelona, Corunna, Seville and other points.
Bombardment of Seville, July 21.
Defeat of Espartero.
1845 Don Carlos assigns his claims to his son.
Isabella II., 13 years old, is declared, by the Cortes, to be of age.
Narvaez, a friend of Queen Christina, is made commander of the army.
1846 Marriage of Queen Isabella to her cousin, Don Francisco d'Assis, Duke of Cadiz.
Marriage of the Infanta to the Duke de Montpensier, son of the King of France.
Protest of England against these marriages.
1847 Attempt by La Riva to assassinate the Queen.
Espartero restored to power.
1848 The British Envoy ordered to quit Madrid within 48 hours.
1850 Birth of the Queen's first child; it dies immediately.
Attempt of Lopez to wrest Cuba from Spain.
1851 Opening of the Madrid-Aranjuez railway.
1852 Merino, a Franciscan monk, attempts to kill the Queen, and slightly wounds her with a dagger.
1853 Narvaez exiled to Vienna.
1854 Espartero organizes a military insurrection at Saragossa and succeeds in making himself prime minister.
The queen-mother impeached, and compelled to quit Spain.
1855 Death of Don Carlos.
1856 Insurrection at Valencia.
Espartero resigns.
A new cabinet formed, headed by Marshal O'Donnell.
Insurrection in Madrid quelled by the government.
Disbandment of the national guard.
Insurrection at Barcelona and Saragossa quelled by O'Donnell, as Dictator.
O'Donnell forced to resign.
Narvaez is made prime minister.
1857 Birth of the prince royal.
1859 War with Morocco.
O'Donnell commands the army in Africa.
1860 Moors defeated at Tetuan and Guadrimas.
Treaty of peace signed, March 25.
Unsuccessful efforts of Ortega to overthrow the Queen and make the Count de Montemolin king, as Charles VI.
Ortega shot, April 18.
The Emperor Napoleon III. proposes to recognize Spain as a first-class power.
The project abandoned, owing to the refusal of England.
1861 The annexation of St. Domingo to Spain ratified.
Spain joins England and France in the Mexican expedition.
1862 Don Juan de Bourbon renounces his right to the throne.
O'Donnell resigns the premiership.
Insurrection in St. Domingo.
1864 Spanish quarrels with Peru.
General Prim exiled for conspiracy.
1864 Narvaez again becomes prime minister.
He advises the relinquishment of St. Domingo; Queen Isabella refuses.
Christina returns to Spain.
1865 Peace with Peru, which is compelled to pay a heavy indemnity.
Queen Isabella orders the sale of the crown lands, and gives three-fourths to the nation.
Spain relinquishes St. Domingo.
Quarrel with Chili, followed by war.
Kingdom of Italy recognized by Spain; insurrection, headed by General Prim.
1866 General Prim lays down his arms, and insurgents enter Portugal.
O'Donnell resigns, and Narvaez forms a new ministry.
The Cortes dismissed by the Queen.
Spain formally recognizes and forms a treaty with the republics of Guatemala, Honduras, Salvador, Costa Rica and Nicaragua.
1867 Revolt in Catalonia and Aragon suppressed.
1868 The Queen grants general amnesty.
Death of Narvaez.
Bravillo becomes prime minister.
Revolution led by Prim and Serrano, Sept. 17; revolution successful, and ministry resigns.
Queen Isabella takes refuge in France, and is deposed.
Provisional government organized at Madrid, by Prim, Serrano and Olozaga, Oct. 8.
Religious freedom, liberty of the press, and universal suffrage granted by new government, Oct. 25.
Revolts at different points suppressed.
The United States government recognizes the provisional government.
1869 Efforts to find a king for Spain.
Serrano elected Regent, June 18.
Prim becomes prime minister.
Outbreaks of the Carlists and republicans suppressed.
1870 Espartero declines the Spanish crown.
Isabella abdicates in favor of her son Alfonso; it is offered to Prince Leopold, of Germany, who refuses it.
Amadeus, son of the King of Italy, elected king by the Cortes, Nov. 16.
Amadeus lands at Carthagena, Dec. 30.
Marshal Prim assassinated, Dec. 28.
1871 Amadeus enters Madrid, Jan. 2.
Serrano forms a new ministry, Jan. 4.
The Cortes dissolved, Nov. 28.
Insurrection in Cuba.
1872 Resignation of the ministry.
Carlist war begins.
Serrano enters Navarre; defeats the Carlists at Oroquieta.
Attempt to assassinate the King and Queen, July 19.
Suppression of Carlist and republican uprisings.
1873 Abdication of King Amadeus.
Republic proclaimed.
Defeat of the Carlists at various points.
Don Carlos enters Spain, July 15.
Cadiz surrenders to him, July 31.
Castelar President of the Cortes.
The "Virginius" affair.
1874 Coup d'Etat.
Marshal Serrano President and Commander of the army.
Overthrow of the republic.
Alfonso XIII. proclaimed king by troops, Dec. 30.
1875 King Alfonso lands at Barcelona, Jan. 9.
Vittoria taken from Carlists, July 9.
1876 Surrender of Bilboa, Feb. 1.
Defeat of Carlists at Durango, and surrender at Pamplona, Feb. 28.
Don Carlos flees to France.
Triumphal entry of Alfonso into Madrid.
1877 Extradition treaty with the United States.
General amnesty to Carlists.
Queen Isabella visits Spain.
1878 Marriage of King Alfonso to Mercedes, daughter of the Duc de Montpensier, Jan. 23.
Death of Queen Mercedes, June 26.
Attempted assassination of Alfonso, Oct. 25.
1879 Inundations in Seville, Granada and elsewhere.
Alfonso marries the Archduchess Maria Christina, of Austria, Nov. 29.
Attempted assassination of king and queen, Dec. 30.
1880 Law for gradual abolition of slavery in Cuba, Feb. 18.
Execution of the assassin Otero, April 14.
1881 Expulsion of Don Carlos from France, July 17.
1882 Franco-Spanish commercial treaty approved by the Cortes, April 25.
Introduction of a bill to abolish slavery in Cuba, June 18.
Heavy snow storm at Madrid, Dec. 18.
1883 Marriage of Infanta della Paz to Prince Louis, of Bavaria, April 2.
King Alfonso visits Frankfort to witness German military maneuvers, Sept. 20.
King Alfonso appointed commander of the Schleswig-Holstein Uhlan regiment by German Emperor, Sept. 23.
Return of Alfonso to Madrid, Oct. 2.
Resignation of Spanish ministry, Oct. 13.
Herrera becomes Prime Minister.
1884 Severe earthquakes in Spain; over 1,000 lives lost, Dec. 25-28.
1885 Resignation of the ministry, in consequence of the determination of the king to visit cholera-stricken districts, June 26.
Terrible ravages of cholera in Valencia and other points.
Spain greatly excited over the occupation of the Caroline Islands by Germany.
Announcement that of [illegible] persons attacked by cholera [illegible] had died, Aug. 21.
1886 Alfonso XIII. King, with Maria Christina as Regent, May 17.
1891 Reciprocity between Cuba and the United States, May.
1892 Riotous demonstrations of Republicans suppressed by the police.
Cargo of dynamite explodes at Santander, killing and wounding several hundreds of people.
1895 Cuban patriots rise again in arms to free their native land. Marshal Campos sent with a large army to suppress the insurrection.

FRANCE.

1769 Beginning of the power of Madame du Barry.
1770 The Dauphin marries Marie Antoinette, of Austria.
1774 Death of Louis XV.; accession of Louis XVI.
1776 Dismissal of Turgot from office.
1777 Necker becomes Minister of Finance.
1781 Necker resigns as Minister of Finance.
The torture abolished in legal proceedings.
1783 Treaty of Versailles; peace with England and Spain.
1785 "Diamond necklace affair" occasions intense excitement.
1787 Meeting of the Assembly of Notables; controversy over taxes.
1788 The Second Assembly of Notables.
Reappointment of Necker.
1789 Meeting of the States-General, May 5.
The Deputies of the Tiers Etat organize themselves as the National Assembly, June 17.
1789 Destruction of the Bastile, July 14.
The beginning of the French revolution.
The king and queen compelled by a mob, at Versailles, to go to Paris, Oct. 6.
The National Assembly meets at Paris, Oct. 19.
The National Assembly changes the royal title to "King of the French," Oct. 16.
Clerical property confiscated.
The division of France into 83 departments, Dec. 22.
1790 King Louis accepts the work of the revolution, Feb. 4.
Titles of honor and hereditary nobility abolished.
Confederation of the Champs de Mars; the King takes the oath to the constitution, July 14.
1791 Flight of the king and queen from Paris, June 20.
Imprisonment of the king and queen in the Tuileries; they are arrested at Varennes, June 21.
Louis sanctions the National constitution, Sept. 13.
Dissolution of the National Assembly, Sept. 29.
1792 First coalition against France.
Commencement of the great wars.
War with Austria declared April 20.
Battle of Valmy; the Prussians defeated, and France saved from invasion, Sept. 20.
Attack and capture of the Tuileries by a mob; the royal family imprisoned in the Temple, Aug. 10.
Massacre in the prisons of Paris, Sept. 2-6.
Opening of the National Convention, Sept. 17.
The Convention abolishes royalty, Sept. 21.
Meeting of the Legislative Assembly, Oct. 1.
France declared a republic, Sept. 22.
Trial and condemnation of King Louis, Nov. 13 to Dec. 18.
1793 Louis XVI. beheaded, Jan. 21.
War against England, Spain and Holland, declared Feb. 1.
Insurrection in La Vendee begins, March.
Proscription of the Girondists.
Robespierre becomes Dictator March 23.
Beginning of the Reign of Terror, May 31.
Charlotte Corday assassinates Marat, July 13.
Execution of Marie Antoinette, Oct. 16.
Siege of Toulon; first victory of Bonaparte.
The Duke of Orleans, Philippe Egalite, beheaded, Nov. 6.
Madame Roland executed, Nov. 8.
Vendee revolt suppressed, Dec. 12.
1794 Danton and others guillotined, April 5.
Elizabeth, sister of Louis XVI., executed.
Robespierre becomes president, June.
Fall of Robespierre, July 27.
Robespierre, St. Just and seventy others guillotined, July 28.
Close of the Reign of Terror.
1795 The Dauphin (Louis XVII) dies in prison.
Napoleon suppresses rebellion of royalists Oct. 5.
The Directory established Nov. 1.
1796 Bonaparte wins the victories of Montenotte, April 12; Mondovi, April 22, and Lodi, May 10; Altenkirchen, June 4; Rastadt, July 5, in Italy.
The conspiracy of Babeuf suppressed.
1797 Pichegru's conspiracy fails.
Return of Napoleon into Paris.
Bonaparte's Egyptian expedition embarks.
Battle of the Pyramids, July 13-21.
Destruction of the French fleet, near Alexandria, by Nelson, Aug. 1.
1799 England, Germany, Russia, Turkey, Portugal and Naples coalesce against Napoleon, June 22.
Bonaparte returns from Egypt; deposes the Council of Five Hundred, Nov. 10, and Napoleon is declared First Consul Dec. 24.
1800 Battle of Marengo, June 14.
Great victory by Bonaparte over the Austrians.
Attempt to kill the Consul by means of an infernal machine, Dec. 24.
1801 Treaty with Germany.
The Rhine made the French boundary.
Peace with Russia, Oct. 8, and with Turkey, Oct. 9.
1802 Defeat of the French at Aboukir, March 8.
Peace with England, Spain and Holland signed at Amiens, March 27.
Legion of Honor instituted.
Bonaparte made "Consul for Life." Aug. 2.
1803 Bank of France established.
War with England declared, May 22.
1804 Conspiracy of Moreau and Pichegru against Bonaparte fails.
Execution of the Duke d'Enghien, March 21.
The empire formed and Napoleon proclaimed Emperor, May 18.
Crowned by the Pope, Dec. 2.
1805 Napoleon crowned King of Italy, May 26.
Destruction of the French fleet, Oct. 21, by Nelson at the battle of Trafalgar.
Battle of Austerlitz.
Austria totally defeated, Dec. 2.
Treaty of Presburg, Dec. 26.
1806 Confederation of the Rhine ratified at Paris, July 12.
Fourth coalition of the Great Powers against France; Prussia declares war, Oct. 8.
Defeat of the Prussians at Jena, Oct. 14.
Capture of Erfurt by the French, Oct. 16.
1807 Russians defeated at battle of Eylau, Feb. 8.
Alexander and Napoleon meet at Tilsit, June 25.
Treaty of peace signed, July 7.
The Milan decree published, Dec. 17.
1808 New nobility of France created.
The beginning of the Peninsular war.
Abdication of Charles IV. of Spain.
1809 Napoleon defeated at Aspern and Essling.
Victorious at Wagram.
Entry of Napoleon into Vienna, May.
Treaty of Vienna, Oct. 14.
Divorce of the Empress Josephine, Dec. 15.
1810 Napoleon marries Maria Louise of Austria, April 1.
Union of Holland with France.
1811 Birth of the King of Rome, afterward Napoleon II.
1812 War declared with Russia.
Napoleon invades Russia.
Great victory of the French at Borodino, Sept. 7.
Disastrous retreat of the French from Moscow, October.
1813 The Concordat treaty with the Pope.
Alliance of Austria, Russia and Prussia against Napoleon, March 16.
Battle of Leipsic.
Napoleon defeated, Oct. 16-18.
The Allies invade France from the Rhine; the English from Spain, under Wellington, Oct. 7.
1814 Surrender of Paris to the Allies, March 31.
Abdication of Napoleon I. in favor of his son, Napoleon II., April 4.
Napoleon goes to the Island of Elba, May 4.
Louis XVIII. enters Paris, May 3.
The Bourbon dynasty restored.
The Constitutional Charter established, June 4-10.
1815 Napoleon leaves Elba and lands at Cannes, March 1, and proceeds to Paris, where he is joined by all the army.
Louis XVIII. leaves Paris; restoration of the empire.
The Allies form a league for his destruction, March 25.

ANCIENT, MEDIEVAL AND MODERN HISTORY.

1815 Napoleon abolishes the slave trade, March 29.
Leaves Paris for the army, June 12.
He invades Belgium, June 15.
Final overthrow of Napoleon at battle of Waterloo, June 18.
Napoleon reaches Paris, June 20.
Abdicates in favor of his son, June 22.
He reaches Rochefort, where he intends to embark for America, July 3.
Entry of Louis XVIII. into Paris, July 8.
Napoleon goes on board the "Bellerophon" and claims the "hospitality" of England, July 15.
Upon reaching England he is transferred to the "Northumberland," and sent a prisoner to St. Helena, Aug. 8, where he arrives Oct. 15.
Execution of Marshal Ney, Dec. 7.
1816 The family of Napoleon forever excluded from the throne of France.
1820 Assassination of the Duke de Berri, Feb. 13.
1821 Death of Napoleon I. at St. Helena, May 5.
1824 Death of Louis XVIII., Sept. 16.
Charles X. becomes king.
1827 National Guard disbanded.
War with Algiers.
Serious riots in Paris.
Seventy-six new peers created.
1829 The Polignac administration organized.
1830 Chamber of Deputies dissolved, May 16.
Capture of Algiers by the French, July 5.
Revolution and barricade of streets in Paris, July 27.
Flight and abdication of Charles X., July 31.
Unpopular ordinances passed regarding the election of deputies and the press, July 26.

[illegible]

Austria-Hungary.

[illegible]

Venetia ceded to France, July 4, and intervention requested.
Great victory by the Austrian fleet over the Italian fleet, at Lissa, July 20.
An armistice agreed upon between Austria and Prussia, July 22; peace of Nikolsburg, Aug. 26.
Hanover, Hesse-Cassel, Nassau and Frankfort gained by Prussia.
Austria retires from the German Confederation.
Baron Von Beust made prime minister.
The Emperor makes great concessions to Galicia.
1867 A new and very liberal Constitution for the empire adopted.
Hungary established an independent kingdom.
Andrassy elected President of Hungarian Diet.
The Emperor and Empress of Austria crowned King and Queen of Hungary, at Pesth, June 8.
1868 The clergy of the Roman Catholic church made amenable to the civil law.
Civil marriage authorized.
The State assumed the control of secular education.
1869 Serious outbreaks in Dalmatia against conscription.
1870 The Concordat repealed.
Neutrality declared in the Franco-Prussian war.
Bitter contest between national and federal parties.
1873 Further reforms in the government instituted.
Measures adopted looking to the representation of all the nationalities embraced in the empire.
Austria recognizes new German Confederation.

[illegible]

SCANDINAVIA.

[illegible]

ANCIENT, MEDIEVAL AND MODERN HISTORY.

GERMANY.

PRUSSIA.

Great Britain and Ireland

ANCIENT, MEDIEVAL AND MODERN HISTORY.

1862 Great distress in the cotton manufacturing districts in consequence of the civil war in America.
Confederate "Alabama" sails from England.
Second international exhibition, May 1.
Marriage of Princess Alice to Louis of Hesse, July 1.
Prince Alfred declines the throne of Greece, Oct. 23.
Serious riots in Ireland.
1863 Continued distress in cotton districts.
Marriage of the Prince of Wales to Princess Alexandra, of Denmark, March 10.
1864 Birth of a son to the Prince of Wales.
Visit of Garibaldi.
The Ionian Islands ceded to Greece.
Fenian troubles in Ireland; arrest of James Stephens, "Head Center," Nov. 11.
Russell-Gladstone ministry.
Death of Richard Cobden, April 2.
Death of Lord Palmerston, Oct. 18.
1866 Defeat of Lord Russell's reform bill, June 18.
Resignation of Russell ministry, June 26.
Derby forms his third cabinet, July 6.
Cattle plague continues, causing great loss.
Princess Helena marries Prince Christian of Schleswig-Holstein, July 5.
Atlantic cable pronounced a success.
Habeas Corpus suspended in Ireland.
Fenian invasion of Canada.

AUSTRALIA.

CANADA.

UNITED STATES.

ANCIENT, MEDIEVAL AND MODERN HISTORY.

ANCIENT, MEDIEVAL AND MODERN HISTORY.

ANCIENT, MEDIEVAL AND MODERN HISTORY.

1876 Opening of the Centennial Exhibition at Philadelphia, May 10; it closes, Nov. 10.
Serious difficulties between Americans and Chinese in California.
Bursting of reservoir at Worcester, Mass., destroying millions of dollars worth of property, March 3.
Death of Alexander T. Stewart, April 10.
War with Sitting Bull and the Sioux.
Massacre at Hamburg, S. C., June.
Massacre of Gen. Custer and his command, by the Sioux Indians, July 2.
Completion of the First One Hundred Years of American Independence; great rejoicing throughout the United States, July 4.
Castle Garden, N. Y., destroyed by fire, July 9.
Younger Brothers and Northfield Bank robbery, Sept. 7.
Arrest of W. M. Tweed, at Vigo, Spain, Sept. 6.
Yellow fever in Georgia, September.
Trial of Molly Maguires, October.
Dastardly attempt to rob the grave of President Lincoln, Nov. 7.
Burning of the Brooklyn Theater, 278 lives lost, Dec. 5.
First furnace for cremation built, at Washington, Penn., Dec. 6.
The Ashtabula railroad horror, Dec. 29.
1877 Close of the Indian War.
The Electoral Commission Bill passed by Congress, Jan. 25, 26.
Rutherford B. Hayes declared President, March 2.
Blue Glass mania.
Death of Cornelius Vanderbilt, June 4.
Great Railroad riots, East and West, July and August.
1878 Yellow fever epidemic along the Lower Mississippi.
Meeting of the Alabama Claims Commission, Feb. 27.
Fenians attempt a second invasion of Canada, May 25.
Death of Robert Dale Owen, June 24.
The Colorado Petrified Giant humbug.
Return of Henry M. Stanley from African explorations, August.
Death of Brigham Young, Aug. 29.
Death of Oliver P. Morton, Nov. 1.
Earthquake shocks in New England and middle states.
Ku-Klux Bill passed by Congress.
Death of Benjamin F. Wade, March 2.
Development of the telephone and phonograph.
Bankrupt Repeal Bill passed, May 15.
Death of William Cullen Bryant, June 12.
Indian outbreak in Washington Territory, July.
Chinese Embassy visits the United States.
Silver Bill passed by both Houses of Congress.
Yellow fever in the South.
Gold sold at par—the first time since 1862—Dec. 17.
1879 Resumption of specie payments, Jan. 1.
Death of Richard Henry Dana, Feb. 2.
Great fire at Reno, Nev., March 2.
New Constitution of California adopted, May 7.
Death of William Lloyd Garrison, May 24.
Terrible tornado in Kansas, Nebraska and Missouri, May 30.
Bill to erect a monument on site of Washington's birthplace, passes both Houses, June 10.
Waterspout in Black Hills causes great loss of property and life, June 12.
Disastrous storms east and west, July.
Great fire at Deadwood, Dak., Sept. 26.
Death of Gen. Joseph Hooker, Oct. 31.
Death of Zachary Chandler, Oct. 31.
Caleb Cushing dies at Madrid.
"Exodus" of negroes from South to West.
James Russell Lowell made Minister to England.
Fall elections favor Republicans.
1880 Death of Frank Leslie, Jan. 10.
City Hall, Albany, destroyed by fire, Feb. 10.
Terrific tornado sweeps over parts of Western and Southern States, April 8.
Great forest fires in Southern New Jersey, April and May.
Collision on Long Island Sound destroys the steamers "Narragansett" and "Stonington".
Centennial celebration of the capture of Andre, Sept. 23.
Garfield and Arthur nominated by Chicago Republican Convention, June 8; Hancock and English by Cincinnati Democratic Convention.
At the General Election, the Republican candidates secured 214 out of 369 electoral votes, Nov. 2.
1881 Electoral College vote counted, Feb. 9.
Three per cent, funding bill passed, March 3.
Steamer Corwin sails for the Arctic regions in search of the Jeannette, March 4.
Revised New Testament issued, May 20.
Star route frauds exposed, May 28.
The great comets of 1881 first seen, June 23.
Sitting Bull, Chief of the Sioux, surrenders, July 21.
James A. Garfield inaugurated, March 4.
Contest between Garfield and Senator Conkling (N. Y.) about New York Collectorship, May.
Commercial treaty with China signed, May 5.
Great Britain pays £15,000 award for damage done to American fisheries in Fortune Bay affair.
Assassination of President Garfield by Charles J. Guiteau, at Baltimore railway depot in Washington, July 2.
Death of President Garfield at Elberon, N. J., Sept. 19, buried at Cleveland, Sept. 26.
Vice President Arthur becomes President, Sept. 20.
Special session of the Senate, Oct. 10.
The celebrated Guiteau trial begins, Nov. 14.
News of destruction of Jeannette, Arctic exploring vessel, Dec. 20.
1882 Guiteau convicted Jan. 25; sentenced Feb. 4; hanged June 30.
Anti-Chinese bill (twenty years) passed March 23; vetoed by the President April 4.
Senate passes Edmunds Anti-Polygamy Bill, Feb. 16; approved March 22.
Apportionment bill passes the House, Feb. 17.
Great Mississippi overflow, wide destruction and loss of life.
Tariff Commission Bill passes both Houses, May 4-9; approved May 15.
Bill extending National Bank charters passed both Houses, May 18.
Violent cyclone at Grinnell, Ia., June 8.
Second Anti-Chinese bill (ten years) passed; signed by President Arthur, May 6.
Collision of the Scioto on Ohio river, 58 persons drowned, July 4.
River and Harbor Bill passed over the President's veto, Aug. 2.
Return of the survivors of the North Pole expedition.
Star Route trial ended by verdict of Jury, Sept. 11, acquitting Turner, convicting Miner and Rerdell, and disagreeing as to Brady, the Dorsey brothers, and Vail.
Steamer Asia founders on Lake Huron, 100 lives lost, Sept. 14.
Utah Commission completes registration of voters, Sept.

1882 The Pendleton Civil Service Bill passes Senate, Dec. 27.
1883 Civil Service Reform bill passes the House, Jan. 4.
Presidential Succession Bill passed Senate, Jan. 9; not considered in the House.
Burning of Newhall House, Milwaukee, 80 lives lost, Jan. 10.
Great flood in Ohio River, 50,000 people homeless, Feb. 10-14.
Tariff and Tax Amendment Bill passes both Houses, March 2.
Death of Alexander H. Stephens, aged 71, March 4.
Death of Peter Cooper, aged 92, April 4.
Cyclone at Beauregard, Miss., 53 lives lost; tornadoes in Iowa and Georgia, April 22.
Opening of the Brooklyn Suspension Bridge, May 24.
Pendleton Civil Service Act passes both Houses, July 16.
Steamer Proteus of the Greely Relief Expedition crushed by ice in Smith's Sound, July 23.
Terrific tornado at Rochester, Minn., many lives lost, Aug. 21.
Northern Pacific Railroad formally opened, Sept. 8.
Civil Rights Act of March 1, 1875, declared unconstitutional by U. S. Supreme Court, Oct. 15.
Gen. Sherman relinquishes command of the army, Nov. 1; Gen. Sheridan succeeding.
Two-cent letter postage goes into effect throughout the United States, Oct. 1.
Serious riot at Danville, Va., between negroes and white military, Nov. 3.
Dakota adopted a constitution erecting Southern Dakota into a State, Nov. 6.
Festivals in honor of the 400th anniversary of Luther's birth, Nov. 10-11.
48th Congress organized.
1884 House repeals the iron-clad oath law, Jan. 21.
Germany returns resolutions of the House laudatory of Lasker, Feb. 15.
United States Supreme Court affirms the constitutionality of Legal Tender Act, March 3.
Mexican War pension bill passes House, March 3.
The Senate ratifies commercial treaty with Mexico, March 11.
Defeat of Morrison Tariff bill, May 6.
Congress appropriates $1,000,000 for New Orleans Exposition, May 8.
Great panic in Wall street; Failure of Grant and Ward and others, May 6-14.
Relief expedition rescues survivors of the Greely Arctic expedition, at Cape Sabine, June 22.
President vetoes the Fitz-John Porter bill, July 2.
Corner-stone of the Bartholdi Statue of Liberty laid, Aug. 5.
The general election resulted in the election of Grover Cleveland, who carried 20 States, securing 219 electoral votes against 182 for James G. Blaine, Nov. 4.
Opening of the 48th Congress, Dec. 1.
1885 Grover Cleveland resigns the New York governorship, Jan. 6.
Dedication of the Washington monument, the tallest structure known, 555 feet, Feb. 21.
Occupation of Aspinwall, S. A., by United States troops.
Inauguration of Grover Cleveland as President, March 4.
New Orleans Exposition opened, Dec. 16.
Treaty with Colombian Government, providing a joint protectorate over the Isthmus, May 6.
The Revised Old Testament and complete Bible published, May 18.
Death of Gen. U. S. Grant, at Mt. McGregor, N. Y., aged 63, July 23.
Grant memorial services held at Westminster Abbey, London, Aug. 4.
Death of Vice-President T. A. Hendricks, aged 66, Nov. 25.
1886 The Presidential succession act signed Jan. 19.
Controversy between the Senate and President over reasons for removing public officers, Jan. 25.
400 Chinamen driven from Seattle, W. Ter., by a mob, Feb. 8.
Death of General Winfield Scott Hancock, aged 61, Feb. 9.
Blair Educational bill passes the Senate, March 5.
Bill for free and unlimited coinage of silver defeated, April 8.
Chicago Anarchist riot, 6 police killed and 61 wounded, May 4.
Anarchists indicted at Chicago, May 27.
President Cleveland married to Miss Frances Folsom, June 2.
Oleomargarine bill passes the Senate, June 25.
Morrison Tariff bill defeated, June 17.
House of Representatives passed bill repealing the pre-emption, timber culture and desert-land laws, June 7.
Bill to repeal the Civil Service law indefinitely postponed by the U. S. Senate, June 15.
Congress requires the Treasury to issue small denomination silver certificates, July 24.
The President warns office holders against attempts to control political movements, July.
Death of Samuel J. Tilden, aged 74, Aug. 4.
Chicago anarchists to the number of 8, found guilty of murder, Aug. 20.
Earthquake at Charleston, S. C., destroying $5,000,000 worth of property and 27 lives, Aug. 30-31.
Surrender of the Apache chief Geronimo and his band, Sept. 4.
Death of Ex-President Chester A. Arthur, aged 56.
Bill to regulate the counting of electoral votes passed, Dec. 9.
1887 Inter-State Commerce bill signed, Feb. 4.
House defeats the Dependent Soldier Pension Bill, Feb. 24.
Retaliation bill passed, March 3.
Bill to redeem trade dollars passed, March 13.
Inter-State Commerce commission appointed, March 22.
Mormon convention at Salt Lake City adopt a Constitution, July 1.
Defeat of the Scotch cutter Thistle by the American Volunteer in race for "America cup," Sept. 27 and 30.
President and Mrs. Cleveland leave Washington for a Western trip.
Mormon convention of monogamists petition Congress for admission of Utah as a State, Oct. 3.
United States Supreme Court refuses to interfere with the finding of Illinois courts in anarchist cases, Nov. 2.
Governor Oglesby commutes death sentences of Schwab and Fielden to life imprisonment, Nov. 10.
Hanging, at Chicago, of the anarchists Parsons, Spies, Engel and Fischer, Nov. 11.
Republican National Committee select Chicago for National Convention, June 19, 1888, Dec. 8.
1888 Terrific blizzard in Minnesota, Dakota and Iowa; 200 lives lost, Jan. 12.
Inter-State Commission confirmed by the U. S. Senate, Jan. 16.
Fisheries treaty with Great Britain signed at Washington, Feb. 15.
Strike of engineers and firemen on the C., B. & Q. R. R. began Feb. 27.

1888 Deadlock in the House of Representatives over the Direct Tax bill, April 5.
Death of Chief Justice Morrison R. Waite, aged 72 years, March 23.
Knights of Labor appeal to Congress for a system of Government telegraph, April 14.
Death of Roscoe Conkling, ex-U. S. Senator, aged 60 years, April 18.
Daily sales of U. S. bonds began, April 23.
Melville W. Fuller, of Illinois, nominated by the President as Chief Justice, April 30; confirmed by the Senate, July 20.
Chinese Treaty ratified by U. S. Senate, May 7.
Execution of murderers by electricity after Jan. 1, 1889, passes N. Y. Senate, May 8; approved by the Governor, June 4.
The President approves of bill to invite a conference of American States at Washington in 1889, May 24.
Lieut.-Gen. Philip H. Sheridan confirmed as General of the Army, June 1.
National Democratic Convention at St. Louis renominates President Cleveland, June 6.
National Department of Labor bill approved by the President, June 13.
The President signed the Chinese Exclusion Bill, forbidding any Chinese laborer who has been, or may now be, or may hereafter be, a resident within the U. S., and may depart therefrom, and who may not have returned before the passage of this act, to return to, or remain in, the U. S., Oct. 1.
Death of General Philip H. Sheridan, aged 57 years, August 5.
Major-Gen. John M. Schofield appointed to the command of the army, August 14.
U. S. Senate rejects the Fisheries treaty, August 21.
President's message to the U. S. Senate recommending enlarged powers under the Retaliation act, August 23.
Floods at Augusta, Ga., destroyed $1,500,000 worth of property, Sept. 11.
Bill prohibiting coming of Chinese laborers approved, Sept. 13.
September wheat touched $2 on Chicago Board of Trade, Sept. 29.
U. S. Supreme Court sustains the constitutionality of the Iowa "Prohibitory Law," Oct. 22.
The "Murchison" decoy letter to Lord Sackville West made public, Oct. 24.
Lord Sackville West, British Minister, dismissed by the President, Oct. 30.
National Election for President; the Republican candidates elected, Nov. 6.
Official yellow fever bulletin gave total number of deaths 412, and of cases 4,705, at Jacksonville, Fla., Dec. 10.
U. S. men-of-war Galena and Yantic sailed for Hayti to demand release of the Haytien Republic, Dec. 12.
1889 Great storm in Pennsylvania; many lives lost at Pittsburgh and Reading, Jan. 9.
Niagara Suspension Bridge blown down at 3 a. m., Jan. 10.
Department of Agriculture created, Feb. 1.
The States of North and South Dakota, Montana and Washington, created by Congress, Feb. 20.
Benjamin Harrison inaugurated President, March 4.
Oklahoma proclamation issued, May 27.
Opening of the Oklahoma country, April 22.
Centennial of Washington's inauguration, April 30.
Murder of Dr. Cronin at Chicago, May 4.
Destruction by flood of Johnstown, Pa.; 5,000 to 10,000 lives lost; over $10,000,000 worth of property destroyed, May 31.
Judge D. S. Terry shot by U. S. Marshal Nagle, defending Justice Field, Aug. 14.
International Marine Congress meets at Washington, Oct. 16.
North and South Dakota admitted by proclamation, Nov. 2.
Trial of Cronin suspects began Aug. 30, ended Dec. 16. Coughlin, Sullivan and Burke found guilty, and received life sentences; Kunze, imprisonment three years; Beggs found not guilty.
David J. Brewer appointed a Supreme Court Justice, Dec. 1.
Death of Jefferson Davis, late President of the Confederate States, Dec. 6.
1890 Appointment of Special World's Fair Committee, Jan. 18.
La Grippe or influenza prevalent throughout the Northern and Western States.
Death of Gen. Crook, at Chicago, March 19.
Act approved providing for the World's Columbian Exposition, at Chicago, April 25.
Death of Gen. Fremont, at New York City, July 13.
First execution by electricity, at Auburn, N. Y., Wm. Kemmler, Aug. 6.
First legislature of Oklahoma meets, Aug. 29.
Act forbidding the use of the mails for lottery purposes, approved Sept. 19.
The McKinley tariff bill takes effect, Oct. 6.
General election; next House of Representatives Democratic, Nov. 4.
The 51st Congress convenes, Dec. 1.
Sitting Bull and seven other Indians killed near Standing Rock Agency, Dec. 15.
Battle of Wounded Knee, between the 7th Cavalry and hostile Indians, Dec. 29.
1891 Death of George Bancroft, historian, at Washington, Jan. 17.
Death of Wm. Windom at a banquet in New York, Jan. 29.
International Monetary Congress met at Washington, Jan. 7.
Application before the U. S. Supreme Court for a prohibition to the U. S. District Court on its decision in the Behring Sea difficulty by Canadian representatives, Jan. 12.
Sioux Indian war ended by submission of the Hostiles, Jan. 15.
Reciprocity treaty with Brazil announced, Feb. 5.
Death of Admiral David D. Porter, at Washington, Feb. 13.
Death of Gen. Wm. T. Sherman, at Washington, Feb. 14.
Charles Foster, of Ohio, appointed Secretary of the Treasury, Feb. 21.
Copyright bill passed Congress, March 3.
Act creating Circuit Court of Appeals, passed March 3.
French Spoliation Bill passed, March 3.
The Copyright bill becomes a law, March 4.
The Enlistment of Indians in the U. S. army authorized March 5.
Proposed arbitration of Behring Sea dispute, March 13.
Lynching of 11 Italians at New Orleans, March 14.
Nicaragua Canal Party sails, March 11.
American Society of Authors formed for the protection of writers, March 26.
Recall of the Italian Minister, Baron Fava, March 31.
25th anniversary of the founding of the Grand Army of the Republic, April 6.
Ground broken for the Grant Monument, New York City, April 27.

1891 Chinese Government refuses to receive the American Minister, H. W. Blair, April 23.
Fort Berthold Reservation, N. D., opened for settlement, May 20.
"The People's Party" formed at Cincinnati, May 20.
Statue of Abraham Lincoln unveiled at Lincoln Park, Chicago, May 22.
Bronze statue of General Grant, at Galena, Ill., unveiled, June 3.
The Czar of Russia presents Stanford University with a complete collection of Russian and Siberian minerals, June 12.
Surrender of the Chilian ship, Itata, at Iquique, to the U. S., June 4.
First shipment of block tin from California mines, June 16.
International Postal Congress held at Vienna decides to hold next Congress at Washington, June 22.
Commercial treaty with Spain signed, June 26.
Transfer of the Weather Bureau to the Agricultural Department, June 30.
$500,000 accepted from the Itata for violation of the U. S. Navigation laws, July.
Libel filed against the arms and ammunition on the Itata, at San Diego, July 12.
Statue of Stonewall Jackson unveiled at Lexington, Va., July 21.
Smokeless powder used for the first time by the U. S. Government, July 25.
The "Majestic" breaks the ocean record, time being 5d. 18h. 8m., Aug. 5.
Cherokee strip in Indian Territory closed to Whites, Aug. 13.
Rain-making experiment at Midland, Texas, Aug. 10.
The "Teutonic" breaks the trans-Atlantic record of the "Majestic," time 5d. 16h. 31m., Aug. 19.
Indian lands of Oklahoma opened, Sept. 22.
Dedication of Pope Leo XIII. statue, presented to the Catholic University at Washington, Sept. 23.
Leland Stanford, Jr., University at Palo Alto, Cal., opened, Oct. 1.
Equestrian statue of General Grant at Lincoln Park, Chicago, unveiled, Oct. 7.
Commercial treaty with Germany concluded, Oct. 11.
Shoshone and Arapahoe Indians sell one million acres of land to the Government at 85 cents an acre, Oct. 15.
U. S. Government demands reparation from Chili for assault on the crew of the Baltimore, Oct. 26.
Argument in the Sayward case, to test U. S. jurisdiction over Behring Sea, begun in the U. S. Supreme Court, Nov. 3.
Congress met; Mr. Crisp, of Georgia, chosen Speaker, Dec. 7.
1892 Stevens County, Kan., war again breaks out, Jan. 5.
Inter-State Commerce Commission appointed by the President, Jan. 5.
Terrible mine explosion at McAlester, Ind. Ter., nearly 100 lives lost, Jan. 7.
Secretary Blaine notifies foreign countries of retaliatory measures, as required by the Tariff Law, Jan. 8.
Special message to Congress from the President, recommending financial aid to the World's Columbian Exhibition, Feb. 24.
The President submits correspondence with England to Congress, regarding Behring Sea controversy, March 8.
Ex-Congressman W. R. Morrison selected as President of the Inter-State Commerce Commission, vice Judge Cooley, resigned, March 31.
Free Silver coinage debate in Congress, March 22-24.
French Extradition Treaty signed, March 25.
The Silver bill shelved, March 25.
The Free Wool bill passed, April 7.
Diplomatic intercourse with Italy renewed, April 11.
Sisseton Reservation, S. D., opened, April 15.
Revenue steamers ordered to Behring Sea, April 18.
Copyright agreement with Germany signed, April 16.
The President approves Behring Sea modus vivendi, April 19.
U. S. Commercial Treaty between Switzerland and Italy, signed April 11.
The President invites foreign nations to participate in an International Silver Conference, April 21.
The President lays Grant monument corner stone, New York City, April 27.
Chinese Exclusion bill signed, May 5.
Terrible floods in the Mississippi Valley, May 5-15.
Wyoming appoints women to National Republican Convention, May 7.
The Alliance party proposes a new currency, May 8.
The Pope approves Archbishop Ireland's Educational Policy, May 10.
Association of American authors formed, May 17.
Reciprocity with Guatemala goes into effect, May 30.
James G. Blaine resigns as Secretary of State, June 4.
Republican National Convention held, June 7.
Benjamin Harrison and Whitelaw Reid nominated, June 10.
Democratic National Convention held, June 21.
Grover Cleveland and Adlai Stevenson nominated, June 23.
Peary Arctic relief expedition sails, June 27.
Homestead, Pa., Steel Works closed, June 30.
Prohibitionists nominate John Bidwell for President, July 1.
People's Party nominate James B. Weaver for President, July 4.
Slaughter of Pinkerton men at Homestead, July 6.
National Christian Endeavor Society Convention at New York, July 7.
Pennsylvania troops take possession of Homestead, Pa., July 12.
Bill to close the World's Fair on Sunday passes both Houses, July 14.
Great storms in Minnesota, July 20.
The President proclaims Oct. 21 a National holiday, July 21.
H. C. Frick, chairman Carnegie Steel Co., shot by Berkman, July 23.
George Shiras confirmed by the Senate as Associate Justice U. S. Supreme Court, July 26.
Inman Steamer City of Paris breaks the Ocean Record, 5d. 15h. 58m., July 27.
Central Labor Union rejects anarchistic resolutions, July 30.
Congress appropriates $2,500,000 to the World's Fair, Aug. 5.
Chinese sailors forbidden employment on American ships, Aug. 4.
International Monetary representatives appointed by the President, Aug. 7.
Trouble among East Tennessee miners, Aug. 12.
Railroad strike of switchmen at Buffalo, great destruction of property, Aug. 13.
The President proclaims retaliation against Canada on canals, Aug. 20.
Nancy Hanks again breaks the trotting record, 2.05¼, Aug. 31.
Death of George William Curtis, author and journalist, Aug. 31.
Cholera brought to New York City by Hamburg steamer Moravia, Aug. 31.
Nelson beats the stallion record, 2.10½, Aug. 31.

1892 Death of J. G. Whittier, poet, Sept. 7.
Nancy Hanks again breaks the trotting record, 2.04, Sept. 28.
Formal opening of the Chicago University, Oct. 1.
Dedication of the World's Fair buildings, at Chicago, Oct. 21.
Fire at Milwaukee destroys 315 buildings, with $5,000,000 loss.
Anarchist monument dedicated at Waldheim Cemetery, near Chicago, Nov. 6.
Great strike at Homestead, Pa., declared off, Nov. 20.
Stamboul lowers stallion record at Stockton, Cal., 2.07½, Nov. 23.
Death of Jay Gould, capitalist, Dec. 2.
Dr. McGlynn restored as a priest, Dec. 23.
Immense gold fields discovered in Utah, Dec. 27.
Prof. Briggs acquitted of heresy, Dec. 29.
Great floods in California, Dec. 25.
George W. Vanderbilt gives a costly art gallery to the Fine Arts Society at New York, Dec. 30.
1893 Death of General Benjamin F. Butler, Jan. 11.
Senate passes the Seal Protection Bill, Jan. 12.
Death of ex-President R. B. Hayes, Jan. 17.
Hawaiian Provisional Government proclaimed, supported by U. S. authorities, Jan. 17.
Death of James G. Blaine, statesman, Jan. 27.
Russian Extradition Treaty confirmed, Feb. 6.
Conflict of rival Legislatures in Kansas, Feb. 11-25.
Rank of American Ambassador established, March 1.
Inauguration of President Cleveland, March 4.
Behring Sea arbitration opened at Paris, France, April 10.
President Cleveland opens World's Fair at Chicago, May 1.
Chinese Exclusion Act goes into effect, May 5.
Governor Altgeld pardons Chicago anarchists, June 26.
Extra session of Congress called June 30.
Great fire at World's Fair, 24 lives lost, July 10.
Behring Sea arbitrators award in favor of England, Aug. 15.
Great storm on South Atlantic coast, Aug. 28.
Wabash railroad accident at Kingsbury, 14 killed, 41 wounded, Sept. 22.
Chicago Day at the World's Fair, attendance 716,881, Oct. 9.
World's Fair closed at Chicago, Oct. 30.
Repeal of the Silver Purchase Clause Act of 1890, Nov. 1.
1894 New York Court of Appeals decides that foreign corporations may hold real estate in New York State, Jan. 16.
Wilson Tariff Bill and Income Tax passes the House, Jan. 31.
U. S. Warship Kearsarge, famous as the destroyer of the Confederate Alabama, wrecked on Roncador Reef, Feb. 7.
Death of George W. Childs, philanthropist and journalist, at Philadelphia, Feb. 3.
Greater New York bill signed by the Governor, Feb. 28.
President Cleveland vetoes the Bland Silver bill, March 29.
Behring Sea proclamation issued, April 10.
Unconstitutionality of the South Carolina Dispensary law declared, April 19.
125,000 coal miners ordered to strike in Ohio, April 20.
Coxey's army invaded Washington, D. C., April 29.
Dr. Talmage's Tabernacle in Brooklyn destroyed by fire, May 13.
177 buildings burned by fire at Boston, May 15.
American Railway Union boycotts Pullman Car Company. Affected 50,000 miles of railroad, June 26.
Armor-plate frauds detected, June 29.
U. S. Court enjoins strikers from interfering with railroad trains, July 2.
Railroad mobs destroy property in and near Chicago, July 6-10.
Railroad strike declared off, July 13.
Utah Enabling Act signed, July 17.
American marines landed at Seoul, Corea, July 25.
Work resumed at Pullman, Ill., Aug. 2.
Hawaiian Republic officially recognized, Aug. 9.
65 factories close at Fall River, 20,000 men idle, Aug. 13.
United States recognizes the sovereignty of Nicaragua over the Mosquito Coast, Aug. 26.
New Tariff becomes a law, without the President's signature, Aug. 27.
Earthquake with great loss of life at Uvalde, Texas, Aug. 31.
Reciprocity Treaty with Cuba cancelled by Spain, Sept. 2.
President Cleveland's Hawaiian letter first published, Sept. 5.
Amnesty granted polygamists in Utah, Sept. 27.
Death of Prof. David Swing at Chicago, Oct. 3.
Death of Oliver Wendell Holmes, Oct. 7.
Government offers to arbitrate in the Japan-China war, Nov. 6.
1895 Famous Mora case settled with Spain.
Cotton States Exposition at Atlanta, Ga., opened.